Ross Beveridge

A Politics of Inevitability

VS RESEARCH

Ross Beveridge

A Politics of Inevitability

The Privatisation of the Berlin Water Company, the Global City Discourse, and Governance in 1990s Berlin

VS RESEARCH

Bibliographic information published by the Deutsche Nationalbibliothek
The Deutsche Nationalbibliothek lists this publication in the Deutsche Nationalbibliografie; detailed bibliographic data are available in the Internet at http://dnb.d-nb.de.

Dissertation Newcastle University, 2010

1st Edition 2012

All rights reserved
© VS Verlag für Sozialwissenschaften | Springer Fachmedien Wiesbaden GmbH 2012

Editorial Office: Dorothee Koch | Anita Wilke

VS Verlag für Sozialwissenschaften is a brand of Springer Fachmedien.
Springer Fachmedien is part of Springer Science+Business Media.
www.vs-verlag.de

No part of this publication may be reproduced, stored in a retrieval system or transmitted, in any form or by any means, electronic, mechanical, photocopying, recording, or otherwise, without the prior written permission of the copyright holder.

Registered and/or industrial names, trade names, trade descriptions etc. cited in this publication are part of the law for trade-mark protection and may not be used free in any form or by any means even if this is not specifically marked.

Cover design: KünkelLopka Medienentwicklung, Heidelberg
Printed on acid-free paper
Printed in Germany

ISBN 978-3-531-18219-3

Acknowledgements

I would first like to thank the School of Geography, Politics and Sociology at Newcastle University for providing me with the PhD Studentship that made this book possible. Additionally, I am very grateful to the Postgraduate Directors, particularly Derek Bell, for further extending this financial assistance. My biggest debts at Newcastle are, however, to my supervisors, Tony Zito and Esteban Castro. I would like to thank Tony for his encouragement, wise words and occasional shoves in the right direction. His support over the last ten years has been invaluable. Esteban, I would like to thank for his generosity and openness, as well as his insightful, well-timed observations. I am grateful to both for the productive and good-natured discussions over the last few years. Going further back, I am extremely grateful to Ella Ritchie for providing me with a start in academia and encouraging me to do a PhD. I would also like to thank Simon Guy for all his help in the early stages and his more general support since then.

At the Institute for Regional Development and Structural Planning (IRS) in Germany, I would like, first and foremost, to express my gratitude to the Director, Prof. Dr. Heiderose Kilper, for generously providing me with a 'second home' throughout the PhD. This institutional context, and the support of many individuals within it, was absolutely integral to the completion of the project. I am, in particular, grateful to Tim Moss. He has been a great source of advice and general assistance throughout. As has Matthias Naumann, who also helped me acclimatise to both Berlin and the IRS. I also thank Frank Hüesker, who was a huge help in conducting the fieldwork in 2006-2007, and with whom I have shared many conversations about PhDs, BWB and Berlin.

Additionally, I would like to thank Kimberly Fitch for sending me chapters from her PhD, Richard Freeman for permission to quote his conference paper, Klaus Lanz for background material and Gordon Macleod for his comments and suggestions at the Newcastle POLIS conference. I also thank Anita Wilke at VS Verlag for her help with the editing of the book. Finally, I am extremely grateful to my family, especially my parents, for all their support over the years. In particular, I thank them for not asking too many questions towards the end. I am, however, most heavily indebted to Laura who was patient, optimistic and very kind to the end.

Ross Beveridge

Contents

List of tables

List of acronyms and abbreviations

AG	Aktiengesellschaft
AöR	Anstalt öffentlichen Rechts
BDO	Deutsche Warentreuhand AG
BDU	Bundesverband Deutscher Unternehmensberater eV
BerlBG	Berliner Betriebe Gesetz
BEWAG	Berliner Städtische Elektrizitätswerke Aktiengesellschaft
BWB	Berliner Wasserbetriebe
CBI	Confederation of British Industry
CDU	Christlich Demokratische Union Deutschlands
EC	European Commission
EU	European Union
EU (ETS)	European Union Emissions Trading System
FDP	Freie Demokratische Partei
GASAG	Berliner Gaswerke Aktiengesellschaft
GATS	General Agreement on Trade in Services
HeLaBa	Hessische Landesbank
IMF	International Monetary Fund

NAO	National Audit Office
NGO	Non-governmental organisation
NPM	New Public Management
OFWAT	Water Services Regulation Authority
ÖTV	Gewerkschaft Öffentliche Dienste, Transport und Verkehr
PDS	Partei des Demokratischen Sozialismus
RWE AG	Rheinisch-Westfälisches Elektrizitätswerk Aktiengesellschaft
SPD	Sozialdemokratische Partei Deutschlands
SVZ	Sekundärrohstoff-Verwertungszentrum Schwarze Pumpe GmbH
TPrG	Teilprivatisierungsgesetz
WAB	VEB Wasserversorgung und Abwasserbehandlung
WB	World Bank
WTO	World Trade Organisation

1. Context, themes and strategy

"Truly, we will become the most 'state-of-the-art' city in the Western World."[1]

Wolfgang Branoner,
CDU Economy Senator for Berlin, 1998-2001,
B. Z., 01.03.1999.

"The aim was to change the philosophy of Berlin…Berlin needed to change mentally."

Senior SPD politician in the CDU-SPD Coalition Government
(1996-1999),
in an interview 2007.

"A quick decision about the water company is imperative…Berlin needs money."

Klaus-Rüdiger Landowsky,
Chairman of the Berlin CDU Parliamentary Group, 1991-2001,
Berliner Morgenpost, 04.06.1998.

On the 29th October 1999, the partial privatisation of the Berlin Water Company, Berliner Wasserbetriebe (BWB), was finalised when the city of Berlin entered a public-private partnership with RWE Umwelt AG/Vivendi (now Véolia) S.A /Allianz. This partial privatisation, the result of a lengthy and controversial policy-making process, occurred at the end of a turbulent decade for Berlin. On the same day, the President of the *Rechnungshof* (the German National Audit Office) criticised the city's government for being around $46 Billion (DM 90 Billion) in debt

[1] All quotations from the German press have been translated from German to English by the author, unless otherwise stated.

(Schomaker 1999).[2] In the ten years after the fall of the Berlin Wall, unemployment had risen by 5% to around 15%, while the city's population had steadily dropped.

This was not how things were meant to be. At the start of the 1990s optimism was high. There was a general consensus amongst politicians, academics and business experts in Berlin, Germany and around the world that Berlin's future was written. Despite the anticipated costs and difficulties of bringing the divided city back together, there was a hopefulness about Berlin's potential to become not just a fully functioning city but a flourishing world city or 'global city'. Having been the main stage of the Cold War, Berlin had become the focal point of the reunification of Germany and the 'New' Europe and was set to regain its place amongst the leading cities of the world.

To achieve this Berlin would have to quickly assimilate to the post-Cold War world defined by economic globalisation: the intensification of international trade, increased flows of capital, the growth of international institutions and the overall expansion of the global economy (Duménil and Lévy 2005, 10). To the experts and the government of Berlin this meant that the city's economy would have to be restructured from subsidised manufacturing industry to the service economy. It would have to become 'competitive', an attractive place for multinational companies to invest in and a centre for all things innovative and high-tech. Capturing something of the *Zeitgeist*, Berlin became emblematic of global shifts to free market economies, the reduction of the role of the state and the promotion of the private sector ethos of the 1990s.

Such thinking provided much of the blueprint for dealing with the challenge of bringing together capitalist West Berlin and the socialist economy of the East. Berlin was perceived to have many advantages that would bring it prosperity. In particular, the city was seen as being strategically placed to become a bridge between Western Europe and the emerging democracies and market economies of the former communist countries in Central and Eastern Europe. It was anticipated Berlin would boom as it returned from Cold War isolation and became the capital of Germany again, Europe's largest economy. This optimism was most visible in the commercial property boom of the early 1990s, built on the anticipated rush of businesses and people to the city.

In the midst of these grand aspirations and rising social and economic problems, BWB, the biggest water company in Germany and the largest employer in Berlin (Wuschick 1997), was partially privatised in 1999. After the privatisation of the water companies in England and Wales in 1989, it was the largest privatisation in the European water supply and sanitation (WSS) sector. As elsewhere around the world, commercialisation and privatisation of public companies had become a key

[2] As part of the period under research pre-dates the introduction of the euro in January 1999, the US dollar is used as the standard currency throughout the thesis. 1 billion is understood as 1000, 000 000 (in German: *eine Milliarde*).

strategy in urban governance during the 1990s, with the BWB privatisation following those of the Electricity (BEWAG) and Gas (GASAG) Companies in 1997 and 1998. By the time of the partial privatisation, BWB had been operating for 4 years as a commercial company outside of Berlin. This venture into international water markets was in many ways characteristic of policy-making in the 1990s: the grand rhetoric promising success in global markets was soon undermined by large financial losses, allegations of mismanagement and corruption.

A politics of inevitability

This book provides a detailed analysis of the intricate, complex and often contradictory partial privatisation of BWB. What and who drove this privatisation? It is beyond doubt that both Berlin and BWB had huge debts, but does this fully explain the decision? Did the city only need money as Landowsky, CDU[3] Chairmen of the Berlin and a key politician in 1990s Berlin, stated? Can the decision-making process be boiled down to the simple economics of 'debts = need for money = necessity of privatisation'? How did Berlin and BWB acquire these debts in the first place? Did the government's policy-making contribute to these debts? Why did they spend millions of euros on employing private sector consultants during the privatisation process? Indeed, what was guiding, informing political decision-making in this period? Why, given the economic and financial problems apparent in the city, was the CDU Economic Senator claiming that Berlin would become "the most 'state of the art' city in the Western World"? What about political ideology? What exactly did a senior SPD[4] politician mean by saying that the aim of policy-making at this time was to "change the philosophy of Berlin"? How is this relevant to the partial privatisation of the water company?

To many both at the time and subsequently, the explanation for the privatisation, despite the controversy often typical of water privatisations, was, indeed, relatively straightforward. 'Necessity' was certainly the cornerstone of the Berlin government's argument and it is not an uncommon one in politics. As with other cases of privatisation around the world, particularly during the 1990s, the 'there is no alternative' mantra of the Thatcher government was potent. Politicians often invoke the inevitability of a certain policy given their reading of the prevailing economic, social and political conditions. In June 2010, the UK's Chancellor George Osborne, presented his plans to implement enormous cuts in public spending as 'unavoidable' given the debts the country had. That the government had huge debts is beyond doubt. What was revealing in Osborne's speech was what

[3] Christlich Demokratische Union Deutschlands (CDU)

[4] Sozialdemokratische Partei (SPD)

he identified as the main causes of the debt. Rather than referring to the fact that the previous Labour government had injected billions of pounds into the banking sector to keep it afloat during the financial crisis of 2008–2009, Osborne focused blame on general public spending under the Labour government. "The years of debt and spending make this unavoidable" (BBC website 2010), he said. In the process, he also shifted the focus away from the conduct of the banks and the 'Bailout' which kept many of them in business.

Presenting a policy in such a way is often an effective political tactic in that it directs attention to the apparent faults of political opponents (and/ or the severity of the situation) and away from the normative and ideological assumptions which underpin all policies. It is instrumental to a politics of inevitability. It works to reduce politics to a discussion of 'needs must' based upon the 'hard facts' and politicians to simple executors of inevitable, even if harsh, measures.

The premise of this study is that for many in 1999, the partial privatisation of BWB was indeed commonsensical and, even for many opponents, it was something of an inevitability. However, the analysis steps outside the parameters of this neat, straightforward explanation and assesses how one policy option, amongst others, came to be seen as the necessary and logical option by many. It problematises the 'hard facts' upon which the decision was apparently made, presenting instead an account in which facts can be political constructions, exposed to, as well as shaped - even prompted - by, normative and theoretical assumptions and political strategies. From this perspective, the overall aim is to show how a politics of inevitability emerged and became embedded. It is an account in which global and local dynamics mix; in which neither political agency nor global structures of neo-liberalism determine outcomes. Instead, this politics of inevitability is an untidy, contingent one, where the tension between the general and the specific, between ideology and politicking, and between key local actors and global economic dynamics run through and, perhaps, even characterise the discussion.

Relevance of study

This book arguably represents the first detailed academic study of the BWB partial privatisation in English. There are, in fact, few academic accounts of the privatisation process. In German, Mohajeri (2006) provides a brief overview of the process and some of the consequences of the privatisation process. In English, research revealed three pieces of work: an EU-funded academic project report, 'Watertime' (Lanz and Eitner 2005a; 2005b) and the work of Fitch (2007a; 2007b).[5]

[5] Relevant chapters of the thesis (2007a) provided by the author.

The 'Watertime' report provides a useful overview of the process. From a critical perspective, it highlights key characteristics of the policy process, including the large amount of consultants employed. Fitch's (2007a; 2007b) account emphasises fiscal pressures as the main driver for the privatisation. Political and ideological factors were, in her opinion, secondary. Fitch thus accepts the official account given by the government for the privatisation. Such an acceptance, also found in Monstadt's (2007) discussion of changes in infrastructural management in 1990s Berlin, is in contrast to the critical accounts provided by researchers (Passadakis 2006 and Werle 2004) working for non-governmental organisations (NGOs). They stress the lack of democratic legitimacy in the policy process, allege corruption or the misuse of political power and view the resulting privatisation as a disaster for the city. Though this policy process was undoubtedly problematic these critical accounts are journalistic and, at times, rely on unreferenced sources in their discussion of the policy process.

This study also adopts a critical approach to the partial privatisation but in doing so aims to provide a 'thick' empirical account of how and why privatisation took place (or to provide at least one version of it). The focus here is not on the consequences of the privatisation deal; this has been frequently discussed (see Lanz and Eitner 2005; Passadakis 2006) and is the subject of a study by Hüesker (2011). Instead, this study explores the tensions in the debate and ultimately aims to provide an account of why privatisation emerged as a policy and how it was realised through the political system.

Within the Berlin context, this study is of particular relevance, given the stated aim of *Die Linke* ("The Left" party) in Berlin to reverse a number of privatisations in the city and the continued controversy surrounding BWB and the contents of the confidential contracts in particular[6]. This is then, first and foremost, a study of privatisation in the water sector and should then be of interest to researchers of water governance, particularly those concerned with debates surrounding the changing role of the state. Furthermore, there is a gap in the literature regarding the actualities of achieving privatisation in WSS in the European context. More broadly this study is significant given the continued spread of, and resistance to, privatisation as a policy tool both within Europe and globally.

A further aim of this project is to show how the more mundane and thus usually over-looked sector of infrastructural management, and water services in particular, became wrapped up in the project of re-making Berlin. There has been much research on how the rebuilding and regeneration projects, as well as the debates they provoked in the 1990s, were a means through which the re-imagining and re-fashioning of Berlin was occurring (for an overview see Latham 2006). In

[6] After a public campaign and successful referendum these contracts were made public by the government in 2011. Controversy remains, however, as campaigners claim that not all of the contracts have been released.

particular, how this ambition to 'normalise' the city, to re-create it as a high-tech global city was often in tension with the history and traditions of Berlin. For example, Marcuse has argued that the reconstruction of Berlin should also be seen as an ideological project to transform Berlin into a 'normal' business friendly capital city, thereby erasing its eccentricities and stark reminders of its past (Marcuse 1998).

Given the broader discussion of Berlin in the 1990s presented here, the study makes a contribution to literatures focused on cities, socio-economic re-structuring and the re-definition of the role of public authorities. In particular, it is hoped that the analytical focus on consultants brings a new perspective to the means by which urban spaces are re-configured in these processes. Finally, in broader terms the study is about the reunification and re-development of Berlin in the 1990s. As such the study should be of interest to urban scholars concerned with how global economic processes and neo-liberal discourses have informed post-reunification policy-making strategies.

Analysing a politics of inevitability

No alternatives: neo-liberalism and economic globalisation in the 1990s

Arguing that economic and financial imperatives leave little options in policy-making may not be a new strategy but it is one that has arguably become more prevalent with the rise of neo-liberalism. As Hay (2007, 100pp) has argued, a defining feature of the shift to neo-liberalism has been the presentation of policies not as the 'best' option but as the only option in a context of economic globalisation:

> "Neo-liberalism is defended not as normatively superior to the alternatives, but as the only option – there simply are no alternatives. It is the very condition of economic credibility and competence in an era of globalisation (Hay 2007, 100)".

In the 1990s privatisation and, more broadly, the political message of the neo-liberal New Right were often presented and perceived in increasingly commonsensical terms: that the market and a minimal state were the "only legitimate and viable future" (Held 2006, 222). Reinforced by political commentaries such as Francis Fukuyama's (2006) *The End of History and the Last Man*, which suggested that the collapse of communism was indicative of the triumph of economic and political liberalism, the message of the neo-liberal New Right became increasingly difficult to ignore (Held 2006, 222). In fact, neo-liberalism became the dominant means through which economic globalisation was both interpreted and promoted.

Left-of-centre political parties increasingly accepted much of this message about the State and markets. The notions of *Die Neue Mitte* ("The New Centre") in Germany and, more broadly the 'Third Way', captured the fundamental political shifts of the 1990s, as left-of-centre parties sought to realign themselves to the perceived realities of the global economy. The political parties of Gerhard Schroeder, Tony Blair, Bill Clinton and others adapted the neo-liberal message to more social democratic political programmes (Peck and Tickell 2002, 50). Nonetheless, policies aimed at reforming the role of the state, particularly variants of privatisation and deregulation of markets, were fundamental to their political agendas. This increasing acceptance of neo-liberal premises about economic globalisation across the political spectrum, led, in turn, to the growing institutionalisation, normalisation and depoliticisation of neo-liberalism in the 1990s (Hay 2007, 98).[7] With both right- and left-of-centre governments, as well as increasingly powerful international institutions such as the World Bank, promoting and implementing privatisation, its growth in WSS and other sectors was considerable in the 1990s.

The BWB privatisation is then emblematic of the broader political context of the period. The 'inevitability', the apparent lack of alternatives to privatisation, may have appeared plausible given that for many in the 1990s the "hegemony of neo-liberalism seemed undisputed" (Munck 2005, 66). To fully grasp the potency of neo-liberalism, it is necessary to see it as more than just a package of economic policies, such as privatisation, with unintended social and political consequences: it is also "a political rationality that both organizes these policies and reaches beyond the market" (Brown 2005, 38). Neo-liberalism is a political project that aims to re-cast societal realms in the terms of a market-based rationality – in terms of efficiency, competitiveness and entrepreneurialism. Crucially, while aiming to construct the world anew according to a rationality of the market, it suggests that this world of markets already exists – that it is a natural state of affairs (Lemke 2001, 13). Importantly, then, as Brown (2005, 40) states, "neo-liberalism involves a normative rather than ontological claim about the pervasiveness of economic rationality and it advocates the institution building, policies and discourse development appropriate to such a claim".

In the 1990s, it became increasingly difficult to draw a distinction between the neo-liberal normative claims as to how the world should be, and the reality of how it was, precisely because it had been so successfully embedded in institutions, policies and discourses. Thus though neo-liberalism remains only one, particularly reductionist, theory of society and the role of government, it has increasingly re-made society and government in its own terms. Critics of neo-liberalism have

[7] Hay (2007, 97ff) has also emphasised the importance of public choice theory and other rationalist approaches to public management in helping to provide a theoretical justification for neo-liberalism.

certainly argued that neo-liberalism has changed the world, though not in the way that is often claimed. It has been stated that neo-liberal policies have also increased poverty and environmental destruction, whilst undermining democracy (for recent discussions see the edited volumes by Saad-Filho and Johnston 2005 and Heynen *et al.* 2007). Some critics have argued that neo-liberalism is little more than a project to restore class power – one that has successfully concentrated wealth (for example, Duménil and Lévy 2005; Harvey 2005).

Despite the emergence of such criticisms, the central normative claim of neo-liberalism that the world operates according to market rules had increasing substance: neo-liberalism increasingly re-shaped the world according to these market rules. It had a "self-actualizing quality" (Peck and Tickell 2002, 35). The "magic of the market" was increasingly taken as a given (Munck 2005, 497). Of course, this re-making of the world occurred not through magic but through actions: international institutions, as well as national governments, private sector actors, academics and other experts promoted neo-liberalism. The real magic lay in the fact that the more neo-liberal policies and a market rationality were promoted the truer neo-liberal claims about the world appeared. For many there were no alternatives. Even if politicians rejected the normative arguments underpinning neo-liberalism, or discerned the negative political and social 'side-effects', it was increasingly difficult for them to reject neo-liberalism in the context of economic globalisation. If they did, "they ran the danger of not being taken seriously or, much more relevant, being cut off from the flow of international resources" (Centeno 2001, 10). A politics of inevitability emerged, one defined in the terms of the necessities of the market. It was both discursively and materially produced, and it centred on "a deference to global economic forces" (Peck 2004, 394).[8]

A politics of inevitability in Berlin

This study explores some of the ways in which a neo-liberal 'politics of inevitability' developed in 1990s Berlin. For all the particularities of post-reunification Berlin, many of the political practices and policies implemented within the city were characteristic of broader neo-liberal trends in policy-making and the shift to new forms of governance. The analysis centres on the BWB privatisation and more generally WSS management. It provides a critical account of the inevitability surrounding the privatisation and an assessment of the political system during the privatisation process. To gain an understanding of how and why the privatisation

[8] For discussions of the effect of globalisation on policy-making from a political science perspective see: Hay and Rosamond 2002; Watson and Hay 2003.

occurred, the study reveals how neo-liberalism was embedded in institutions, policies and discourse (Brown 2005, 40).

This book problematises the reasons provided for the partial privatisation. It thus problematises the neutrality of logic and facts. Adopting a broad construct-ivist perspective, the emphasis is placed on how meanings are produced in policy–making; how 'facts' are embroiled in discursive strategies; how truths such as the 'no alternative' discourse are constructed in policy-making. So, rather than accepting that it was logical to privatise BWB the project assesses this logic itself, unpacking the assumptions on which this logic was based and revealing in the process how a neo-liberal political rationality shaped the 'facts' within the BWB debate. Rather than seeing the eventual privatisation as commonsensical, the aim is to show how it was that privatisation was justified as the only alternative. With this shift, the partial privatisation of BWB, rather than simply being inevitable, is now seen as being constructed as an inevitability.

Neo-liberalism informed a global city discourse which shaped urban policy-making in Berlin during the 1990s: it set-up necessities in the context of economic globalisation. The study examines some of the broader policies and practices aimed at making Berlin a global city. Neo-liberal assumptions and logics also came to shape WSS policy-making in 1990s Berlin. BWB was re-made, reformed in the terms of neo-liberalism, first as a commercial player in global water markets and then as a partially privatised company. Turning to how this policy was devised and implemented, an attempted re-definition of the policy process as a commercial process was apparent. The study links this 'depoliticisation' of the process to the emergence of new governance arrangements centred on private sector consultants working for the government. Although the privatisation process also raises significant questions about the legitimacy of policy-making, the decision-making arrangements that emerge are still in part restrained and subject to the formal processes of decision-making. Moreover, the effects of neo-liberalism are shown to be contingent.

Structure and agency in policy-making

As stated, by placing the fiscal situation of Berlin in the broader context of the 1990s a fairly straightforward account of the privatisation is possible. Berlin was struggling with the economic and social effects of reunification, economic globalisation was 'forcing' governments at all levels to restructure economies and neo-liberal ideologies and policies rose to prominence as a means of responding to these challenges.

While such points are crucial to the analysis it is necessary not to be deterministic in explaining the interdependencies between them. Privatisation does not just happen and has not occurred everywhere one might expect it to. In his study of the New York City water supply crisis of the 1990s, Gandy (1997) located the crisis within the broader socio-economic restructuring of the city. From the mid-1970s onwards New York experienced severe socio-economic restructuring similar to that observable in 1990s Berlin, as a decline in population, a shrinking industrial base and rising unemployment led to reduction in tax revenue for the city government. For a while in the 1970s the city was entirely dependent on the Federal and State grants (Gandy 1997, 339).

Within this context there was a sharp shift from Fordism to neo-liberalism, which in Gandy's view precipitated both the crisis and a more general transformation of water governance. However, despite disinvestment in infrastructure, privatisation did not emerge as a serious policy option in New York in the 1970s and 1980s (despite its growing use in other parts of the world during this period). Prices may have risen sharply from the late 1980s and the organisation of water supply and sewage services companies may have been reformed, but the reduction in the role and outgoings of the government did not extend to a change of ownership (Gandy 1997, 339). When, in 1995, Mayor Rudy Guiliani did propose full privatisation of the water system as a means of dealing with the city's budget deficit, he was blocked by the city's comptroller who argued that it would provide only a short term solution, whilst risking long term problems for the water supply system (Gandy 1997, 346).

As this example reveals, the actions of individuals or organisations matter. The New York case cautions against seeing privatisation as emerging inevitably from fiscal problems and a general shift towards neo-liberalism. Privatisation is not a natural, normal reaction. Nor is it an inevitable response to the changing needs of the global economy. Globalisation does not have incontestable needs; general shifts to neo-liberalism can be contested and privatisation rejected. There are alternatives to privatisation. It must be achieved and it is therefore contingent upon local, as well as global dynamics.

An emphasis on structural factors risks ignoring the role played by local actors in the BWB partial privatisation. The Berlin political scene is also an object of empirical enquiry, as it was through these actors, and their competing strategies, that Berlin was governed; that the aspiration of the global city was pursued and BWB ultimately privatised. Reared in the heavily subsidised West Berlin of the Cold War, Berlin's political class has been characterised as something of an 'old boys' network where the same faces within the CDU and SPD held power and made decisions (Interviews: 2, 4). Corruption had long been a feature in West Berlin politics and allegations continued in the 1990s (Rose 2004). Perhaps inevitably, allegations of corruption were to dog the large swathe of privatisations in the 1990s (Rose 2004). To some it was of little surprise when the so-called 'Banking Scandal' of 2001 led to the resignation of the Mayor, Eberhard Diepgen and CDU Parliamentary group leader Klaus-Rüdiger Landowsky (two key figures of the so-called 'old-boys' network during the 1990s), the collapse of the CDU-led coalition and the instigation of criminal proceedings against some CDU politicians.

Politics in 1990s Berlin was not, however, dominated only by the 'old boys' of the CDU and SPD. In terms of the privatisation programme, a female, non-Berliner is perhaps the most prominent figure. Annette Fugmann-Heesing (SPD), made Finance Senator in January 1996, was something of an 'outsider', and certainly not a member of the political club in Berlin. Fugmann-Heesing was brought to Berlin by Klaus Böger, the SPD Group Leader and by this stage a supporter of privatisation. Böger wanted to push through a range of reforms including privatisation and Fugmann-Heesing had a track record of implementing such measures in the State of Hessen. Fugmann-Heesing's privatisation programme provoked fierce opposition from within her own party, the *Bündnis '90/Die Grünen* (Alliance '90/ The Greens) and the Former Communist Party, *Partei des Demokratischen Sozialismus* (PDS, now part of the Left Party- *Die Linke*) and even tested the strength of the coalition with the CDU.[9]

The BWB partial privatisation is also a case in which consultants feature prominently. Consultants were prominent in policy-making in general during the 1990s, particularly in the provision of advice as to how Berlin should pursue its global city objective. With regard to the privatisations of the public companies, Fugmann-Heesing assembled a hugely expensive team of private sector consultants at the Finance Senate. This team of legal, tax and financial experts led by Merrill Lynch assumed a central role in the management of the BWB privatisation. Additionally, other consultants played a key role throughout the privatisation process, proposing models to reform BWB, outlining political strategies and lobbying for a range of actors such as the Berlin government (the Senate), private sector companies and even the Unions.

The Union representing the BWB workers, *Die Gewerkschaft öffentliche Dienste, Transport und Verkehr* (ÖTV), were also important political players in Berlin, enjoying close relations with the CDU leadership in particular. As were the Management Board of BWB. Essentially appointed by the Berlin government, BWB was led from 1994-1999 by the prominent ex-CDU Federal politician, Bertram Wieczorek. He had been brought into lead the company in its new 'commercial' phase, to make BWB a global player in the emerging water markets, particularly of Central and Eastern Europe.

[9] *Die Grünen* and PDS from this point on in the thesis.

Policy-making should be seen as the result of a mix of general trends in political thinking and the very particular circumstances of a place. Gandy (1997, 339) states that "to trace the flow of water and examine the discourses surrounding its use, distribution and quality is to illuminate the functioning of urban space in all its complexity and contradictions". WSS policy-making is never simple and the privatisation of BWB should not be seen in such terms either.

In developing an approach to analyse the privatisation, the theoretical approach must take into account both macro and micro dynamics, but it should do so in a way that provides a sense of the interplay between the two. It has to avoid a simplistic understanding of the interdependencies between WSS in Berlin, the particular politics of the city and global discourses about the role of the state. All are connected, but not in a deterministic fashion. Shifting discourses about the role of the state in WSS do not necessarily lead to changes in all contexts of water governance. Furthermore, even when broader trends shape aspects of water governance, outcomes are not predictable. Marrying this appreciation of the effects of broader discourses and local contingencies requires a conceptualisation of the interplay between the macro and micro in policymaking: between discourses, such as neo-liberalism, and political agency.

Policy-making does not, of course, occur in a vacuum. There is a need to consider the institutional setting of policy-making. In analysing the privatisation of BWB, this project rests upon a more fundamental question: how can we best understand and analyse contemporary policy-making processes in representative democracies, which are far from fixed in terms of structures and patterns of governing? This question touches upon a range of crucial debates in the social sciences. First, it is at the heart of concerns about the health of contemporary democracies, the extent to which institutions and decision-making processes fulfil their stated objectives: providing legitimacy, representation and accountability. Second, a concern with the workings of policy-making runs through broader contemporary discussions about governance and governing; that to understand contemporary democracies the key is to trace patterns of governing and reveal the alignments of state and non-state actors.

In Berlin in the 1990s the institutional setting was complicated. The decade began with the integration of the socialist East into a West Berlin political system, still largely characterised by West German "neo-corporatism" (Schmitter 1974): the institutionalisation of class interests in policy-making (labour Unions and employers) with economic and social policy largely defined through collective bargaining between the government, Unions and business (Streeck 2006, 14). However, as the decade progressed, politics became increasingly defined by shifts to new forms of neo-liberal governance. The traditional, fixed structures of policy-making became less prevalent as relationships between the state, civil society and the private sector were newly articulated (Swyngedouw 2005, 1994).

Any analysis of policy must, then, address the philosophies of state-society relations, the exercise of political power and the process of policy-making itself (Hill 1997, 41; Wilson 2000, 247). As a key tool in the re-drawing of the role of the state, an examination of privatisation must address the specific arguments put forward for privatisation as well as more general notions about how the state should change. Further questions are also apparent. How should the relationships between facts about the world and political philosophies be understood? How should the relationships between the broader phenomenon of neo-liberalism and local political agency be conceived? How does inevitability in policy-making emerge? How did the political system function within this neo-liberal context? What institutional forms emerged in the privatisation process? And what implications did this have for the accountability, transparency and legitimacy of policy-making?

To address these issues the study argues in favour of a move away from the positivist approaches that have dominated policy studies. The approach adopted is based on readings from three literatures: the post-positivist policy studies literature; Neo-Foucauldian governmentality studies; and, from science and technology studies, Actor-network Theory (ANT). The following section outlines how this approach is developed and utilised for the analysis of the BWB partial privatisation.

Outline of the book

Chapter 2 places the case study in context. It begins with an overview of privatisation, its rise in the 1980s and 1990s and provides a summary of a period increasingly defined by neo-liberalism and globalisation. Policy-making in 1990s Berlin is also placed within this context: how the apparent necessities of globalisation and the means of neo-liberalism shaped the competition between cities encapsulated in the notion of the global city. The chapter then turns to privatisation in the WSS sector in this period. In particular, the discussion focuses on the main arguments presented for privatisation, namely that privatisation brings greater efficiency. Although this was not the main argument put forward by the Berlin government for privatisation, the notion itself has been fundamental to the broader rise of privatisation. Taken with the complementary argument that the private sector brings 'know-how', it apparently provides evidence that privatisation brings with it a superior form of management. These claims have been made by a wide range of actors: governments, financial institutions (such as the World Bank and IMF), public water companies themselves (e.g. the BWB Management Board), private sector consultants and academics. This chapter shows that such claims have been increasingly contested by academic studies: that the 'facts' of privatisation are far from simple.

Chapter 3 presents a discussion about claims in politics: claims about the reality of things; the nature of relations between facts and values in policy-making; and the claims made about decision-making in political systems. It is argued that mainstream positivist policy studies do not provide a convincing account of how facts and values mix in policy-making. In particular, such studies, in their pursuit of 'impartial' analyses, through the testing of hypotheses and the employment of methods from the natural sciences, provide little sense of the ways in which policy-making rests upon the construction of truths. In particular, such studies have a tendency to reify the object of study and ignore the discursive dimension. This is not to argue that there is no 'evidence' which can be seen to support the claim that privatisation has in certain cases led to some discernable improvements. Rather it is to stress that the process of producing evidence must be destabilised and shown to be contingent.

From this perspective the Berlin government's presentation of the BWB privatisation in such simple and seemingly neutral terms is at best problematic and at worst misleading. It obscures the biases inherent in the policy: the desire to expand the market and reduce the power of the state. Privatisation is always a *political* decision and never merely an 'evidence-based' decision or a case of 'needs-must'. The presentation of privatisation in such terms is not unique to Berlin. This leads into a broader discussion of issues researchers have raised concerning the political institutions of representative democracies. In particular, the claims to 'rationalism' and 'objectivism' which underpin formal institutional policy-making processes (Dryzek 1994). These claims are crucial to the democratic legitimacy of policy-making in liberal democracies but they can also serve to conceal the extent to which 'facts' are contestable – that they are based on assumptions and strategies of action. Experts are highlighted as being central to this production of facts. In particular, with the rise of neo-liberalism, private sector consultants have become more prominent in policy-making. They are, as stated, crucial to the BWB partial privatisation and their presence in government brings into question the traditional roles of political institutions.

Chapter 4 outlines the analytical approach, bringing together insights from governmentality studies, Actor-network Theory and Hajer's discourse analysis approach to policy studies. Following the work of key writers in the field of governmentality studies (Andrew Barry, Mitchell Dean, Thomas Lemke, Wendy Larner and especially Peter Miller and Nikolas Rose) governing is seen as a problematising activity informed by particular political rationalities (e.g. liberalism, Keynesianism and neo-liberalism). The notion of 'governmentality' provides a starting point for thinking about government and water policy-making. It brings attention to the inter-dependencies between thought and government; highlighting "how governing involves particular representations, knowledges and often expertise regarding that which is to be governed" (Larner and Walters 2004, 2).

In particular, the study turns to the key work of Rose and Miller (1992), *Governing beyond the state: problematics of government.*[10] They proposed three dimensions to the study of government. First, governing, the projection of political power, should be analysed in terms of the fundamental contests between "political rationalities" such as neo-liberalism and Keynesianism. These are the broader discursive frames through which different forms of political power are rationalised (Rose and Miller 1992, 175). They do not, of course, simply become political realities. They need to be realised through "programmes of government", the substance of political rationalities (Rose and Miller 1992, 181). These programmes problematise aspects of society; they define the scope and duties of government, objectifying aspects of the social and designing the means to transform them (Rose and Miller 1992, 181). Rose and Miller's final dimension to the project of governing, "technologies of government" refers to the means through which political programmes are rolled out in a particular field; the "strategies, techniques, and procedures" by which actors attempt to make their programmes operable (Rose and Miller 1992, 183). From this perspective governmentalities are made possible through forms of expertise. Experts, such as consultants, are crucial to the translation of political rationalities into political practices. The key point is that this is not a neutral process of re-representing reality but one based on the particular theoretical and normative assumptions of political rationalities such as neo-liberalism. From this perspective governing is the "production of 'truths'" about the world (Larner and Walters 2004, 2).

Having outlined a governmentality perspective the chapter then seeks to refine the approach for the study of policy-making, in the process addressing some of the criticisms of the literature that it assumes the effects of discourse on political agency. There is a need to take a closer look at the diverse means through which actors, as well as intellectual, institutional and physical resources are brought together to realise governance in a particular place (Dean 2008, 212). This requires a more detailed assessment of local political discourses, and the interplay between political rationalities such as neo-liberalism and actors in policy processes so as to better reveal the extent to which local factors play a role in shaping how policy was made. Here Maarten Hajer's (1997; 2002; 2003a; 2003b) policy discourse analysis approach provides a way of conceiving of the discursive means by which global discourses such as neo-liberalism were localised in Berlin. Policy discourses, with their "narrative storylines", "policy vocabularies" and "generative metaphors" (together, the "terms of a policy discourse" {Hajer 2003a, 103}) have "won power effects", which shape "the knowing and telling one can do meaningfully" (Hajer

[10] This article was re-published as Miller and Rose (2008) and is also quoted in the thesis.

2003a, 107). Tracing the terms of policy discourses reveals the intellectual and normative processes through which meaning is produced in policy-making.

Ultimately, however, policy has to be made by actors. Therefore the chapter goes on to place an extra emphasis on some of the general points made by ANT writers, such as Michel Callon, Bruno Latour and others and utilised to a certain extent by Miller and Rose. Of specific relevance is the argument that new knowledge is the outcome of relations – that it is produced through the construction of networks of relations (Callon 1986, 203); through the negotiation of mutual objectives between actors (Rutland and Aylett 2008, 633). Translation, as understood here, is more a methodological guide than a theoretical approach. It is a useful means of thinking about the ways in which neo-liberalism was made real by actors. It reminds the researcher that discourses have to be materially constituted: that to be made real, discourses became dependent on the negotiations between actors. From such a perspective, the partial privatisation of BWB can be seen as the outcome of actions and was not simply an effect of a neo-liberal discourse. There is a tension here, of course, one that the study explores but leaves unresolved. Indeed, it shapes the analysis of the BWB case in Chapters 5 and 6.

Chapter 5 argues that a 'global city policy discourse' dominated policy-making in 1990s Berlin. It was a discourse in which the broader material and discursive context (of economic globalisation, the demise of the Soviet bloc, the 'end of history', etc.) was apprehended in Berlin, made sense of in the context of reunification and re-entry into capitalist markets. This policy discourse provided a neo-liberal way of thinking about the tasks facing Berlin's economy and policy-makers. It is argued that decision-making in the water policy sector as well as others was shaped by the translations of this global policy discourse. Indeed, this global policy-making ultimately shaped the discursive and material conditions of the BWB partial privatisation. It 'set-up' the BWB partial privatisation.

This moves the study away from official accounts of the privatisation. Thus this analysis of the partial privatisation of BWB will begin not with a consideration of the financial situation of the Berlin government 1997-1999 – the debts of the city and the costs of reunification at the time of privatisation debate. Nor will it begin with a discussion of the need for private sector expertise and investment in BWB. Instead, it will begin with the vision of a 'New Berlin' as a 'global city'.

Chapter 6 examines how the political system functioned in realising the privatisation. Taking the more formal policy implementation stage of the process (July 1998-October 1999) as its focus, the chapter aims to unearth the actual institutional processes through which the privatisation was realised. Drawing on in-depth interviews with a wide range of participants and press reports from the period the chapter re-constructs the process. To get to the heart of these issues, it is necessary to move beyond official accounts. From piecing together the accounts of the process, it is apparent that a new arrangement of governance emerged to 'manage' key aspects of policy-making and implementation: a consultant-led project team at the Finance Senate. It is argued that formal institutional processes were contested by this Finance Team who attempted to de-politicise the privatisation.

Chapter 7 provides an overall assessment of the case, highlighting the wider significance of the findings. It assesses the means through which neo-liberalism shaped WSS policy-making in Berlin in the 1990s. The chapter outlines the problematic features of the privatisation, locating concluding remarks in a broader discussion of consultants, governance and neo-liberalism.

Chapter 8 concludes the study with some reflections on the theoretical approach employed, summarising the insights it has provided as well as the limitations that have been exposed. Overall, it argues that the approach, with its dual focus on discourse and political agency, has the potential to be applied to other WSS privatisation cases and policy-making more generally. A few examples are provided to illustrate this. The study then closes with a discussion of how privatisation and other fields of contemporary politics might be explored in the future.

Note on methodology

This study employs a case study methodology. A case study is a detailed, empirical inquiry of a phenomenon or practice and the context from which it emerges (Yin 1994, 13 in McNabb 2004, 358). As a strategy for research it has a number of advantages. It allows the researcher to bring together a number of theoretical and empirical elements in order to illuminate that phenomenon or practice (Howarth 2005, 331). The aim of case study research is not necessarily to produce generalisable conclusions or "general context independent theory" but rather context-specific knowledge (Flyvbjerg 2006, 223). In fact, the case study strategy, with its emphasis on detail, aims to expose the contingencies of politics in a particular place at a particular time (Howarth 2005, 331). Providing a 'thick' description of the BWB case entails an engagement with the uniqueness of Berlin: the legacies of its role in the Cold War, the tasks of reunifying a divided city and re-entry to the global economy.

Though there is an emphasis on contingency, wider implications can still be drawn from case studies. The BWB case was selected in part because it is emblematic of prominent trends in 1990s politics: the influence of economic globalisation on policy-making; the rise of neo-liberalism and the shift to governance. It is, in many ways, a "paradigmatic" case, in that it provides some insights into broader trends (Howarth 2000, 331). The BWB case certainly fulfils one of Yin's (1994, 147 in McNabb 2004, 368) measures for successful case studies: that they should be of interest to a wide audience.

As can be seen in Appendix 1, interviews were conducted with a wide a range of participants. There are two obvious gaps: RWE and the CDU. RWE never replied to requests for interviews. Despite numerous efforts, it was not possible to interview senior CDU politicians working at the Economy Senate. This can probably be attributed to the CDU's involvement in the Banking Scandal of 2001, which led to the collapse of the CDU-led coalition government. To balance this, a CDU politician working at the Finance Senate was interviewed. Given the continuing controversy surrounding the partial privatisation, the names of all interviewees have been anonymised.

There are, of course, problems with relying on interviews as the primary source of empirical data, particularly with participants involved in highly controversial policy process that occurred some time ago. Beyond ensuring that there were multiple perspectives on this process, the technique of triangulation (Howarth 2005, 330) was utilised. Findings from interviews were thoroughly cross-checked with secondary source material such as press reports and the academic literature.

2. Privatisation, globalisation and neo-liberalism: governance in the 1990s

Introduction

Although this study is concerned with privatisation in terms of the policy that has emerged since the 1970s, privatisation is not a new phenomenon. It can, for example, be traced back to ancient Greece, where the government owned the land, forests and mines but 'outsourced' work to firms and individuals (Megginson and Netter 2001, 323). With regard to water supply and sanitation services (WSS), private ownership and management were dominant in Europe in the 18th and 19th centuries. The term itself is, however, relatively new and is most commonly associated with the Thatcher government of the 1980s. Perhaps surprisingly, however, its roots are German. Although widely believed to have been coined by the American economist Peter Drucker in the late 1960s, it has been revealed that a form of the term was first in use in Germany between the 1930s and 1950s (Bel 2006, 192). *Privatisierung* emerged as an academic term to describe the Nazi economic policy of transferring some of the state's monopolies to the private sector. One objective of the policy was to consolidate support for the regime amongst the business elite. There is a "rich historical irony" (Bel 2006, 193) here: the contemporary arguments made against privatisation (that it ultimately only benefits political and business elites) are similar to the arguments made in favour of privatisation in 1930s Germany.

Since the 1980s privatisation has transformed the way in which a range of services are provided across the globe. Along with other policies, most notably the liberalisation of markets, privatisation has been a key means of reforming the role of government and re-organising state-society relations. The re-ordering of the roles and responsibilities of public and private actors is vividly illustrated in the water sector. These changes are usually proposed on the grounds that they encourage new forms of flexible governance in which public and private interests coalesce in a positive sum game. This line of thinking has resulted in increasing moves towards various mixes of private and public-private forms of ownership, management and service provision.

Accordingly, contemporary privatisation is difficult to precisely define. Castro (2002, 2) has pointed out that little distinction is made between privatisation in terms of the transferral of ownership to the private sector and private sector participation, which can refer to a multitude of things (such as contracting out services to the private sector). In strict terms, privatisation can be said to cover several "distinct and possibly alternative means of changing the relationships between government and the private sector": the sale of public assets; deregulation, the introduction of competition in monopolies; and sub-contracting or outsourcing of the provision of goods and services (Kay and Thompson 1986, 18).

On the one hand, this variety is tribute to the appetite and innovation of those attempting to implement private sector logics and ownership. On the other, it points to the strength of resistance to privatisation and the consequent reforms and compromises, which have been necessary to implement it. Such is the diffuse nature of private sector influence in terms of ownership and practices, it is clear that thinking of privatisation only, or even primarily, in terms of ownership is misleading. Increasingly, publicly owned companies are operating according to private sector logics framed around competition and efficiency. Indeed, some public companies have been forced to adopt these logics and practices by their public owners. Examples in the water sector include the publicly owned company in Durban (Loftus 2004; 2006 in Swyngedouw 2009b, 39), Scottish Water and, within Eastern Germany, the commercialisation of WSS services in Brandenburg and Mecklenburg-Vorpommern (Naumann 2009, 103, 152, 300).

Such moves have not been without criticism and opposition. Privatisation is usually controversial and always 'political', particularly with regard to WSS services. It might seem strange to emphasise this, given the protests water privatisation has sometimes provoked, but privatisation has at least to some become not only a 'normal', but even an apolitical policy 'tool'. Implemented by governments of both the Left and Right, reflective of an apparently 'post-ideological' politics, privatisation has spread rapidly around the world since the 1980s. Proponents of privatisation often talk or write of it in terms which suggest its merits are virtually self-evident: that it will lead to greater efficiency, better services, higher environmental standards, wider ownership in society through the sale of shares and even counter corruption through reducing the role of the state (Kay and Thompson 1986, 19; Poole 1996, 2-4). In such accounts, the 'political', contestable character of privatisation disappears, obscured in particular by the apparent economic benefits to be had from its implementation. If there is contest and conflict, this is seen to be representative of economically unrealistic objectives, 'outdated' thinking or political dogma. Hence, the 'there is no alternative' mantra of Thatcher's government.

Throughout the 1990s supporters of privatisation presented it as a necessary response of government to economic globalisation: the apparent shift in the organisation of the global economy away from the system centred on and dominated by the traditional state (Swyngedouw 2004, 27). The main manifestations of economic globalisation are the expansion of international trade, increased flows of capital, the growth of international institutions and the global economy itself (Duménil and Lévy 2005, 10). Whether or not economic globalisation is an entirely new phenomenon is contested, as is the extent to which this shift away from the state is being driven simply by 'normal' market forces and advances in technology (for discussions see Cox 1997; Goldblatt *et al.* 1997; Jessop 2000). It has been argued by some that neo-liberal ideology has driven or justified these changes (e.g. Thrift 1999). The role of international institutions (such as the World Bank), the governments of major economies (particularly the USA) and the multinational companies themselves in promoting and imposing economic globalisation has also been stressed (for a discussion see Held and McGrew 2003). What is indisputable, however, is the degree to which the 1980s and in particular the 1990s witnessed significant structural changes in the global economy and a growing ideological consensus in favour of market liberalisation and privatisation.[11]

The structural and discursive effects of globalisation and neo-liberalism were undoubtedly powerful in the 1990s (and continue to be so today). The sense that governments had little room for manoeuvre, little option but to conform to global norms, was captured in the increasing dominance of the 'there is no alternative' (to privatisation or liberalisation of markets, etc.) discourse (Peck and Tickell 2002, 34). The fundamental politico-economic re-alignments in the 1990s are, to some extent, captured by the widespread implementation of privatisation and the increasingly apolitical, commonsensical way in which its supporters presented it. Governments of all hues could be found implementing privatisation in some degree or form, from the communist regime of China to the New Labour government in the UK, who accepted, indeed developed upon, Thatcher's privatisation programme. In fact, governments formed from left-of-centre, social democratic parties, previously opposed to privatisation, became some of the most "successful privatizers" (Poole 1996, 2); Australia, and New Zealand being prime examples.

This chapter first provides an overview of privatisation from the late 1970s onwards. The discussion places privatisation in the context of neo-liberalism, economic globalisation and debates over governance in the 1990s. It reveals how the discourses of neo-liberalism and economic globalisation became entwined, normalising a world defined by markets and the logics of efficiency, competitiveness and entrepreneurialism. It then shows how these discourses have moulded urban governance, leading the government in Berlin to aspire to global city status. Following this discussion of the broader context of the 1990s, the chapter turns to privatisation in WSS services, the arguments used to justify it and the academic

[11] This has often been referred to as the 'Washington Consensus' though originally the term was more directly associated with the reform agenda of international institutions and the USA for developing Latin American countries: see Williamson 2000.

debate that has arisen around it. It concludes by contesting the 'facts' in favour of privatisation and resisting the attempted depoliticisation of debates on the subject.

The rise of contemporary privatisation

The rise of privatisation in the 1980s and 1990s is normally attributed either to the emergence of globalisation or neo-liberalism and the broader contestation of the role of the state in the 1970s and 1980s (Von Weizsaecker 2005, 175). The global economic crisis of the 1970s exerted pressure on government spending and problematised Keynesian, state interventionist policies. There had, as a result of the Depression, World War 2 and widespread decolonisation processes, been a general consensus in most Western industrialised countries that the state should control key utilities and economic sectors such as defence and steel (Megginson and Netter 2001, 325). There were, however, notable exceptions. What has been described as the "first ideologically motivated denationalization programme of the postwar era" was in 1960s West Germany (Megginson *et al.* 1996, 118). The centre-right government of Konrad Adenauer partially privatised state-owned companies such as Volkswagen, which it had inherited from the Prussian state. The stated aim of these privatisations was to increase the popularity of stock ownership amongst the working class, a form of 'popular capitalism' similar in its objectives to Thatcher's government in the UK in that both also aimed to undermine the power of the Unions (Vogelsang 1988, 198). These experiments with privatisation in Germany were, however, short-lived and unsuccessful with the government bailing out many who had invested in shares when the prices collapsed (Vogelsang 1988, 198).

Privatisation in its modern form has been implemented in a range of sectors including water, gas, electricity, coal, steel, banking, insurance, telecommunications and transport. The broad scope of privatisation reveals the extent of change in government in recent times. Since its emergence in Chile, the USA and the UK in the late 1970s and early 1980s, privatisation has been implemented in over 100 countries (Megginson and Netter 2001, 321). In Europe during this period privatisation was most extensive in the UK, but other countries also began to experiment with it. In France the Chirac government privatised 22 companies between 1986 -1988. Outside of Europe, privatisation emerged in Japan in the late 1980s and, more controversially, became very prevalent in Latin America, particularly Mexico. Privatisation did not, however, make major inroads in West Germany during the 1980s. This was despite a move to the right with the election of the conservative-liberal government of the CDU-CSU and the *Freie Demokratische Partei* (FPD). In fact, although this government installed privatisation as a stated political objective (Esser 1998, 101), the only real push for change during this period came externally from the European Commission (EC) (Vogelsang 1988, 196).

The commitment to privatisation in West Germany in this period has been seen as largely "symbolic" (Esser 1998). By the end of the decade public ownership was still dominant in the sectors of telecommunications, postal services, rail and air travel, electricity, gas, water and banking. There was strong opposition to privatisation, particularly from the Unions and the state bureaucracy. More generally, the German economy remained 'organised', with the state playing a strong role. Corporate governance was defined by company structures that limited the control of shareholders and distributed power amongst a wide range of private and public sector actors (including managers, employees and regional authorities) and, of most relevance to WSS, the "sheltered infrastructural sectors" defined by minimal competition and a high degree of state control (Beyer and Hopner 2003, 179). Additionally, there were legal obstacles to privatisation: most public enterprises were public law corporations that legally could not have profit-making functions (Vogelsang 1988, 209).

Neo-liberalism, globalisation and Berlin in the 1990s

The symbolism of Berlin is hard to ignore. The Nazi regime, the end of the Second World War, Communism, the Cold War, the reunification of Germany and Europe, are all associated with the city. In the 1990s the reunification of the city became a 'metaphor' for not merely the reunification of Germany but "that whole period of historical shift" (Cochrane and Passmore 2001, 343). If the breaking down of the Berlin Wall came to symbolise the end of the Cold War, the trends in governance of the city over the next ten years seemed to confirm Francis Fukuyama's (1989/ 2006) 'end of history' thesis that the end of the Cold War marked the "unabashed victory of economic and political liberalism" (Fukuyama 2006, 108). As Fukuyama wrote in 1989, the:

> "victory of liberalism has occurred primarily in the realm of ideas or consciousness and is as yet incomplete in the real or material world. But there are powerful reasons for believing that it is the ideal that will govern the material world in the long run" (Fukuyama 2006, 114).

The battle of ideas had, for many at least, apparently been won. To the government and its advisers, as well as many observers, the task was to make the ideas a reality in Berlin, particularly in economic terms (given the city's virtual disconnection from capitalist markets during the Cold War). Even if there were not necessarily "powerful reasons" for believing that economic liberalism would bring success in

Berlin during the 1990s, the growing dominance of liberalism around the globe was increasingly apparent.

The early 1990s witnessed fundamental shifts in political discourse, especially that concerning the role of the state and markets. In the wake of the collapse of communism, discourses of globalisation and neo-liberalism rose to prominence. It is important to consider the relationships between the two because both are fundamental to the analysis of the BWB partial privatisation and urban governance in 1990s Berlin. The reunification of the city was only part of the story of 1990s governance in Berlin. As stated, the overarching challenge facing the government was to re-integrate the city in the global capitalist economy after more than 40 years of disconnection. East Berlin had been the capital of a socialist economy. West Berlin had been a *Schaufenster* ('display window' or 'showcase') for capitalism, but in reality its declining economy had been propped up by subsidies from the government of West Germany. As Strom (2001, 4) states, "what happens if you insulate a city from most global market forces for years and then suddenly remove barriers to the flow of capital?"

Globalisation and neo-liberalism

Without getting into the debates surrounding what globalisation actually is and whether it is such a new phenomenon, a brief outline is nonetheless required. There is at least general agreement on the core features of the discourse, if not the reality, of globalisation. Generally, it refers to the "expanding scale, growing magnitude, speeding up and deepening impact of interregional flows and patterns of social interaction" (Held and McGrew 2003, 4). Economic globalisation has been identified by Jessop (2000, 83) as referring to the following processes in particular:

1. The internationalisation of national economic spaces through growing inward and outward flows; the establishment of regional economic blocs (such as the EU).
2. The development of economic links between local and regional authorities in different countries.
3. The global scope of multinational companies and transnational banks exploiting favourable conditions in specific localities.
4. Opening of national borders through liberalisation.
5. Widening and deepening of international regimes covering economic activity.
6. The introduction and acceptance of global norms and standards e.g. globally integrated markets and globally-oriented strategies.

While the true significance of these processes is contested, the discourse of globalisation refers to the apparent shift in the organisation of the global economy away from the system centred on and dominated by the traditional state (Swyngedouw 2004, 27). It became a self-evident, virtually incontestable set of arguments and beliefs in the 1990s (Swyngedouw 2004, 27). It was also to become entwined with the discourse of neo-liberalism. Indeed, this occurred to such an extent that "neo-liberal globalisation" has become a term of popular usage (Larner and Walters 2004, 8). Still, it is possible and necessary, given the importance to understanding policy changes towards privatisation in the 1990s, to distinguish between the two discourses regardless of how interdependent they may have become. In doing so, the ways in which neo-liberalism shapes the apparent *requirements* of economic globalisation, whilst providing the policy *means* through which governments should react, becomes evident.

There is general agreement on the core forms of policy practices that fall under the rubric of neo-liberalism. Such is the agreement that, from being the label of choice of critical social science literatures a few years ago (Larner 2000), now even *The Economist* employs the term to describe reform of markets (Leitner *et al.* 2007, p 1). The origins of the term lie in the early Cold War critiques of the then dominant form of Western capitalism (Keynesian or state interventionist government) as a threat to liberty and democracy by Friedrich Hayek and others. Since then, neo-liberalism has manifested itself in:

> "'neo-Schumperterian' economic policies favouring supply-side innovation and competitiveness; decentralisation, devolution, and attrition of political governance; de-regulation and privatisation of industry, land and public services; and replacing welfare with 'workfarist' social policies" (Leitner *et al.* 2007, 1).

Neo-liberalism refers to a "particular political project" (Larner and Walters 2004, 8). It denotes the attempts to reform Keynesian government, replacing state ownership and regulation across a range of economic, social and political realms with the actors and logics of the market. It has taken a number of forms: the state-imposed experiments in Chile in the 1970s (the "proto-neo-liberalisms"); the "roll-back-the-state" projects of the 1980s (as epitomised by the Margaret Thatcher and Ronald Reagan governments); and the "roll-out" projects of the Tony Blair and Bill Clinton governments in the 1990s which aimed to soften the negative social effects of neo-liberalism whilst further embedding market rationales in political and social institutions (Peck and Tickell 2002, 41).

Neo-liberal globalisation

If the internationalisation of trade is not necessarily a new trend (and is, instead, one that was observable in the 19th century {Duménil and Lévy 2005, 10}), what economic globalisation came to mean in the 1990s certainly was. Economic globalisation is commonly interpreted as a "process denoting the universal, boundless and irreversible spread of market imperatives in the reproduction of states and societies across the world" (Colás 2005, 70). It is here that neo-liberalism and economic globalisation can be seen to have merged and reinforced one another. The neo-liberal "vision of a free economy and minimalist state" (Peck and Tickell 2002, 41) has shaped the way in which economic globalisation is framed – the tasks it is seen as providing governments. Economic globalisation *requires* governments to re-draw the boundaries between the state and markets, while neo-liberalism provides a set of policy means with which to achieve this.

By the beginning of the 1990s, in the wake of the collapse of communism, privatisation and de-regulation became the basis of the 'short, sharp, shock' to restructure Central and Eastern European countries. Globally, both policies were also being promoted by key international institutions such as the EU and the World Bank, International Monetary Fund (IMF) and World Trade Organisation (WTO) globally. Within the EU the increasing development of the Single Market (particularly after the agreement of the Single European Act in 1986) has arguably created a regulatory climate conducive to privatisation. The EC and the European Court of Justice had, from the late 1980s, enforced measures to increase market integration and competition in telecommunication, postal services, air and rail transport, energy and water supply sectors (Beyer and Hopner 2003, 179). This push for liberalisation and privatisation continued and intensified in the 1990s, placing ever increasing demands on national governments to comply with the regulations of the Single Market.

As stated, a dramatic shift was also seen in the policies of left-of-centre parties, who, having experienced long periods out of government in countries such as the USA, UK and Germany, came to adopt much of the thinking about the state and policies it advocated. The 'Third Way' of governments led by Bill Clinton, Tony Blair's New Labour and the *Neue Mitte* ('New Centre') of Gerhard Schroeder's SPD in Germany (who came to power in 1998) became – to varying degrees - advocates of privatisation. In the German context, some trade unions even dropped their opposition to privatisation: the German Railway Workers Union (*Eisenbahnergewerk-schaft*), for example.

In Germany the costs of reunification provided an additional argument to those advocating the "downsizing and restructuring" of government (Poole 1996, 1). The development of privatisation has been characterised as "conservative, cautious and pragmatic" (Esser 1998, 102), at least in comparison to other countries such as the UK. Still, the Federal government under Kohl extended privatisation to infrastructural networks where state ownership had previously been seen as "indispensable" (Beyer and Hopner 2003, 186). Increasingly subject to EU legislation regarding competition, telecommunications, postal services and the railways were privatised or put up for discussion. Most privatisations were partial, with the state maintaining a significant share. For example, in 1996 *Deutsche Telekom* was partially privatised with shares being sold for around €10.5 Billion with the state maintaining roughly 30% (Beyer and Hopner 2003, 188). Other privatisations included *Deutsche Lufthansa* and the *Bundesdruckerei.* In total the Federal Government made in the region of €19 Billion between 1994-2000. This occurred within broader reforms of Germany's approach to economic governance, bringing it into line with global economic trends: symbolised, for example, in the move of major companies like Daimler-Benz to "shareholder-oriented" practices (Beyer and Hopner 2003, 180).

'Governance' and depoliticisation

Such neo-liberal political re-alignments have been far from uniform, but Fukuyama's (2006, 114) prediction that politics would become increasingly characterised by consensus, and government by "economic calculation" and the "endless solving of technical problems", resonated through the 1990s. This view of government as an essentially problem-solving activity informed broader theoretical debates about the changing role of the state in the 1990s: that there was a move from government to new forms of 'governance'. Though a contested term, the generally agreed meaning of governance is that governing is increasingly defined by new structures in which the "boundaries between and within public and private sectors have become blurred" (Stoker 1998a, 17). Political decision-making is no longer seen as occurring only through the traditional institutional processes of political systems, but through arrangements including private economic actors and civil society. Further, arrangements of governance do not rely on the formal authority of political institutions, but rather actions have to be coordinated, roles and responsibilities negotiated between public and private actors (Stoker 1998a; Kooiman 2003).

Discussions of governance have often rested on normative assumptions that such new forms of governance are an improvement on traditional government (Mayntz 2003, 32). They are seen to be indicative of a "common purpose, joint action, a framework of shared values, continuous interaction and the wish to achieve collective benefits which cannot be gained by acting independently" (Swyngedouw 2005, 1994). Working from these assumptions, state and international institutions (like the European Union and the World Bank) actively encouraged the development of new forms of governance that allowed market and civil society actors more substantial roles in policy-making, administration and implementation (Swyngedouw 2005, 1992). Forms of governance centred on non-state actors have often been promoted – by the World Bank and the United Nations, for example – as necessary forms of 'depoliticisation', required in building state capacity and market confidence (Flinders and Buller 2006, 294).

Depoliticisation, a move to indirect forms of governing (Flinders and Buller 2006, 295-296), centres in particular on replacing – at least in part – politicians with experts and re-defining political processes in technical terms. As a particular strategy within governance, depoliticisation has been widely promoted by governments as well as international institutions as a means of replacing the political infighting and inefficient bureaucracy of the political process, with the neutrality and effectiveness of de-politicised policy made by technical experts (Hay 2007, 93). One consequence of this trend is that neo-liberal policies, such as privatisation, have increasingly been devised and implemented by technical experts working in public-private arrangements of governance. The World Bank, for example, facilitated and promoted trans-national policy networks, centred on experts, to implement green "neo-liberalism" in developing countries (Goldman 2007).

It can be said that the theories and practices of governance and neo-liberalism have worked to reinforce each other since the 1990s. On the one hand, depoliticisation provides neo-liberal policies with a neutral, apolitical sheen to add to their 'necessity' in the context of globalisation. On the other hand, neo-liberalism, with its emphasis on management and problem-solving to deal with the necessities of economic globalisation, provides the conditions and rationale for the involvement of non-state actors in governing. The "peculiarly depolitising manner of neo-liberalism's consolidation and legitimation" (Hay 2007, 99) has led a number of academics to express concerns about the health of liberal democracies: to argue that neo-liberalism (with its 'necessities') and de-politicising strategies of governance are undermining democracy.[12]

As Flinders and Buller (2006, 296) observed, depoliticisation is something of a misnomer because policy-making is always political. The selection of one course of action and not another will always benefit some more than others – a policy cannot be depoliticised. Rather, policy can be made easier to manage and less controversial. More accurately, depoliticisation refers to a shift in the arena of policy-making, which changes the way in which decisions are made (Flinders and Buller 2006, 296). However, new forms of governance are themselves problematic because private actors such as technical experts lack democratic legitimacy and are not bound by the same rules and obligations as State actors (Mayntz 2003, 34).

[12] For a more detailed discussion of the literature on depoliticisation see: Flinders and Buller 2006; Hay 2007, 90ff).

These criticisms have informed broader critiques of contemporary democracies. The "disembedding of elites" from the formal structures of democracy and the "trivialisation of politics" (the re-definition of politics in organisational and management terms) has been identified by Crouch (2004) as part of the condition of "post-democracy" in advanced liberal democracies. Mouffe (2005) has taken issue with those academics who have, since the end of the Cold War and the rise of globalisation, promoted a "post-political", consensual view of democracy. For Mouffe (2005) the very assumption that a broader consensus exists or can be achieved in democratic politics ignores the antagonistic, conflictual activity necessary for democracy to exist. In a similar vein, Swyngedouw (2007a; 2009a) has written of a "post-politics" reduced to management and expert administration. To Swyngedouw (2007a; 2009a) the depoliticisation of democracy is largely the result of the dominance of neo-liberalism, which is, in turn, further consolidated by the marginalisation of antagonistic views. He has argued that democracy has become an "empty signifier" (Swyngedouw 2007a, 3) in which neo-liberal tendencies have reduced the "political terrain to a post-democratic arrangement of oligarchic policing" (6). This is a politics of apparent consensus, led by public-private administrative elites: a politics of inevitability in which the outcomes of policy-making –what is possible, desirable, who should be included and excluded – are virtually known in advance (Swyngedouw 2007a).

The thrust of these critiques is that the emergence of neo-liberalism and new forms of governance since the 1990s have shaped a politics not of greater democratic opportunities, but one which is increasingly technical in nature, expert-dominated and detached from citizens. As Hay (2007, 122) states:

> "The institutionalisation and normalisation of neo-liberalism in many advanced liberal democracies in recent years have been defended in largely technical terms, and in a manner almost entirely inaccessible to public political scrutiny, contestation and debate".

Neo-liberal globalisation and the city

Within this broader context, the perceived challenges for cities arising from globalisation centred on achieving 'competitiveness' – becoming as attractive as possible to mobile firms looking for the best conditions for conducting business. Accordingly, the workforce should become flexible and knowledge-based, regulation should be kept to a minimum, enterprise should be encouraged and governments should exercise fiscal prudence (Swyngedouw 2004, 27). As will be shown, the Berlin government did what most governments at every scale were doing (Swyngedouw 2004, 27): they adapted policy to the perceived needs of the global economy. Further, they aimed to make Berlin a 'global city'. It was in the pursuit of this objective that the influence of neo-liberal discourse can be discerned.

Urban governance in 1990s Berlin can, in part, be seen as a project of conforming to global norms of urban governance. This study of a privatisation of a water company, located within the context of 'inevitable' socio-economic restructuring of the Berlin economy, is then emblematic of the wider political and economic context of the 1990s. As Peck and Tickell (2002, 34) state, it was a period in which neo-liberalism became the "commonsense of the times". This study does not aim to explain how neo-liberalism became so globally dominant in this period, how it came to have a power which was as "compelling" as it was "intangible" (Peck and Tickell 2002, 34). This is, rather, the backdrop to a study that seeks to explore how neo-liberalism was reproduced in the particularly symbolic context of the reunified Berlin.

After decades of isolation Berlin was thrust back into the global economy. Berlin required a new image, a new self-identity, as well as more concrete means by which the city could be brought together. Though this presented huge challenges to a city whose economy had – on both sides of the wall – been subsidised, many both inside and outside Berlin were confident that the city would prosper in the global economy. Much of this thinking was encapsulated in the image of the 'global city'. Some inspiration for this came from Berlin's past, in particular the Berlin of the 1920s: the 'Golden Twenties', the last time the city had been united, democratic and engaged with the global economy (Eckardt 2004, 2). During this time Berlin had been seen as a 'world city', one that had emerged from the unification of Germany in 1871, and the industrialisation and growth of the city which followed it. The notion of the 'global city' captured the contemporary sense of what a world-leading city was in a context of globalisation and neo-liberalism.

The global city: an aspiration for the 'New Berlin'

The global city notion does, then, rest on the understanding that economic globalisation combined with technological developments, state fiscal crises and neo-liberalism has increased the "place competition" between cities to attract investment (Gleeson and Low 2000, 16). Such trends have led to a shift in the governance of cities, away from 'managerialism' to an 'urban entrepreneurialism' (Harvey 1989). Underlying this shift in the role cities play is the increasing contestation of the nation state in the organisation of global economies. As a result, cities have become more important arenas for the organisation of transnational markets and are increasingly defined according to their functional importance to the operation of global economic networks. Cities are characterised as being in competition with each other to become sites of strategic functions and services in global markets: to be seen as key sites of globalisation or internationalisation (Cochrane and Passmore 2001, 341).

The specific roles cities play in globalisation and how they become seen as key nodes is at the centre of the academic debate on the subject. In perhaps the best known account, Sassen's (2001) "global city" is defined by its concentration of major businesses, especially in the finance sector, which results in it becoming a control centre in global markets. Similarly, "world cities" are locations for the headquarters of multinational companies, where their expertise, management and research functions are concentrated (Beaverstock *et al.* 1999). Alternatively, Castell's (1989 in Cochrane and Passmore 2001, 341) "informational cities", are key points of production, expertise and leadership that arise as a consequence of the emergence of the information based network society.

The emphasis, regardless of the differences, is that cities are in a competitive system: that if the control of world trade has become much more centralised through globalisation, then national and city governments have increasingly acted to ensure their countries and cities remain competitive (Gleeson and Low 2000, 7). Cities do then need to be entrepreneurial. They need to have the reputation of being 'global' due to the connotations it has: of connectedness, dynamism and significance. In this quest for prestige and economic strength, some have argued that the thinking behind the global city is not that new, but is instead a "very traditional notion of a high ranking service metropolis with various international connections" (Kraetke 2001, 1779). Regardless of how these developments should be interpreted, the argument that cities have to adapt themselves to the global economy has become almost a truism to many politicians, administrators and those advising them.

> "Not only is the existence of globalisation and of world cities widely accepted as the common-sense of the age, but the implicit – and often explicit – agenda for urban development is how best to take advantage of this perceived new world"(Cochrane and Passmore 2001, 341).

Such is the orthodoxy of this view that even cities that have no chance of becoming global leaders continue to play the game, to compete on the same business-friendly terms. This is certainly the path that many in Berlin have pursued as a means of somehow combining "one dream – that of national capital-building – with another – the promise of becoming a global city" (Cochrane and Passmore 2001, 343). The key point to be taken from this discussion is that an understanding of urban governance and policy-making in the water sector must be located within an appreciation of the broader discursive context of globalisation and neo-liberalism and how these shaped a 'storyline' in Berlin politics:

> "Berlin is re-defining itself and being re-defined within the constraints of 21st century urban development discourse and global political economy" (Cochrane and Passmore 2001, 343).

The point to stress is that, by following this global city path, other possible paths for Berlin were lost and previous patterns of governing in Berlin forgotten. Trends in West Berlin politics suggested Berlin might have developed rather differently than it did in the 1990s. For example, one means employed to tackle high unemployment in the 1970s and 1980s was developing partnerships with societal organisations. Such community-based governance was complemented by democratic checks in planning projects; the high priority given to green issues; and the focus on, and encouragement of, small businesses (Bruegel 1993, 156). All this appeared to offer a "distinct base for the development of the post-reunification Berlin economy" (Bruegel 1993, 156). In 1989, there was even a "short-lived hope that Berlin (and East Germany in general) could go down a 'third path' between western capitalism and German Democratic Socialism" (Campbell 1999, 178). As will be shown, the government often followed a different path during the 1990s, one which reflected more general trends in urban governance: developing public-private partnerships to develop the economy (Harvey 1989) and playing 'catch-up' in urban planning as private investment re-shaped the urban landscape in the property boom.

Nonetheless it is worth noting that the image of Berlin as an up-and-coming global city did not convince everyone. An early set-back was the unsuccessful attempt to merge Berlin with the surrounding region and state, Land Brandenburg. In a referendum on the subject in 1996, the majority of voters in East Berlin and Brandenburg voted against it, rejecting the vision of a global or European city-region put forward by politicians and private sector urban marketing firms (Cochrane and Passmore 2001, 345). Such failures did not, however, lead the government of Berlin to consider alternatives. As will be shown, the global city aspiration remained intact throughout the 1990s despite mounting economic problems.

Privatisation in the water sector

Privatisation in the WSS sector gradually emerged within the more general problematisation of Keynesian policies from the 1970s onwards and the consequent moves to neo-liberal policies. Although Thatcher's government privatised the water companies in England and Wales in 1989, privatisation of WSS services has generally developed more slowly than in other sectors and been more controversial and problematic in its effects. This can be seen as a reflection of both the inherent obstacles to implementing private sector logics – of profit-making and competition - in the natural monopolies that usually characterise WSS sectors as well as the special status and value of water as a fundamental resource for living and an environmental treasure.

However, as in other sectors, private sector involvement was justified on the grounds that it would bring much needed investment and efficiency, the former seen as particularly crucial in the developing world (Hall *et al.* 2005, 286). Privatisation became a big business, with some $468 Billion being raised from the sale of public assets and functions to the private sector between 1984-1994 and $80 Billion in 1994 alone (Poole 1996, 1). Privatisation would continue to increase until the end of the 1990s, with global revenues topping $140 Billion every year from 1997 until 2000, when they reached $180 Billion (Megginson 2005, 20). Coming in the last half of the 1990s, and culminating with the privatisation of BWB, Berlin's privatisations (of utilities, banking, landholdings and housing) occurred at the peak of the privatisation movement.

Privatisation is usually promoted as a means of introducing 'market discipline' and achieving 'efficiency gains' and is justified with reference to public choice theories, property rights theories and neo-classical economics (Castro 2002; Megginson and Netter 2001, 329). Writing in the mid-1990s as privatisation became a favoured policy of governments, another academic supportive of privatisation wrote of it in terms of the "downsizing and restructuring" of government (Poole 1996, 1). The sense in this description is that government could be understood in the same terms as business; that privatisation could be seen as a 'managerial' decision based on the sound economics of government.

Though arguments for privatisation have rested largely, but not solely, on these 'economic' benefits, close inspection of both the theory and practice of privatisation have revealed inconsistencies. Moreover, there is no clear empirical evidence to suggest that private sector involvement leads to improved performances and greater efficiency – though much of the mainstream academic literature might suggest otherwise. Typical would be the concluding remarks of an article assessing the economic performance of privatised companies:

> "We know that privatisation 'works' in the sense that divested firms almost always become more efficient, more profitable, and financially healthier and increase their capital investment spending" (Megginson and Netter 2001, 381).

Such claims are countered by other studies (for detailed discussions see Castro 2002; Hall and Lobina 2005). For example, Letza *et al.* (2004) have attacked the "myth" that privatisation equals greater efficiency, and have argued instead that efficiency has as much to do with social and commercial factors as ownership. Though such critical appraisals of privatisation have increased in recent years (matching its growth around the world), even in the 1980s academics expressed doubt over whether private ownership brought about superior performance. A survey (Kay and Thompson 1986) of studies of public and private ownership conducted in the mid-1980s concluded that there was no evidence that private sector performance was "intrinsically superior" (23). There were, quite simply, efficient and inefficient public enterprises and efficient and inefficient private firms (24). Furthermore, the authors argued that where there was evidence of superior private sector performance, this was generally found in competitive markets and not the natural monopolies of gas, water and electricity (25).

The experiences of privatisation in the water sector during the 1990s and 2000s confirm such conclusions. Even arch-promoter of privatisation, the World Bank, has questioned the benefits of privatisation (Castro 2002, 4-5; Hall *et al.* 2005, 287). In recent years the world's largest water companies, Suez and Veolia Water (formerly Vivendi), have pulled out of some developing countries complaining about high costs, risks and problems associated with public opposition (Hall *et al.* 2005, 286; Swyngedouw 2009b, 41). The most notable examples of public resistance have occurred in the developing world. Perhaps the most famous example is the Cochabamba 'water wars' in 2000 that resulted in the international consortium of water companies losing their contract. Other successful campaigns were carried out in Buenos Aires and Tucuman in Argentina against Suez, though similar campaigns in Chile and the Philippines proved less successful (Hall *et al.* 2005, 286).

Water privatisation in Europe, though rarely provoking violence and civil unrest on the scale found in Bolivia, has also rarely been a popular decision. In the UK for example, there was little public enthusiasm for water privatisation. Opposition between 1985-1987 persuaded the Thatcher government to shelve plans for water privatisation (Hall *et al.* 2005, 287), while continued opposition in Scotland prevented privatisation when it was eventually implemented in England and Wales in 1989 and public disenchantment with privatisation played a role in the mutualisation of Welsh Water in 2000. Water privatisation was even rejected during the 'short, sharp, shock' of restructuring in 1990s Central and Eastern Europe. Debrecen in Hungary (1994) and Lodz in Poland (1994) both rejected privatisation in policy debates. More recently the Polish city council of Poznan rejected privatisation in 2002 (Hall *et al.* 2005, 288).

Within the German context, privatisation of water services has remained limited: Rostock in 1992, Potsdam in 1998 and Berlin in 1999. Water services in the German context have remained largely in the control of the municipalities, a reflection of Germany's distinct model of water and sanitation services (Barraque 2009, 241). Early in the 20th century *Stadtwerk* appeared in German cities as an alternative to centralisation and privatisation: the merging of a city's utilities and public transport services into one company (in more recent times sewage collection and treatment, solid waste services and cable TV have also been merged). Barraque (2009, 214) has argued that this reflects German trust in municipal government.

Certainly the experiences of the partially privatised water companies as well as other privatisation debates in German cities seem to confirm this observation. The Potsdam privatisation was reversed in 2000 (Hall *et al.* 2005, 288), while Munich rejected privatisation in favour of maintaining public control in 1998, as did Hamburg in 2005. In Berlin, there is a citizens' group campaign for re-municipalisation. *Die Linke*, junior partner in the Coalition government with the SPD, is formally committed to reversing the partial privatisation. The Berlin SPD themselves, despite their prominent role in the privatisations of the 1990s and 2000s, are now willing to consider re-municipalisation of BWB. Resistance to privatisation in Germany can also be found at the highest levels of government. In 2003, the German Federal Government and many *Länder* Parliaments passed motions opposing any attempted moves by the European Commission (EC) under the GATS (General Agreement on Trade in Services) of the World Trade Organisation (WTO) that could lead to the opening up of water sector to foreign private sector competition.

Elsewhere there have been moves towards re-municipalisation. In Europe, the Mayor of Paris announced in 2008 that the contract with the private water company would not be renewed when it expired in 2009. This continued a trend in France toward re-municipalisation (e.g. Grenoble) while, elsewhere around the world, privatisation was reversed in cities like Atlanta in the USA, where the partnership with Suez-owned subsidiary, United Water, was cancelled by the city government in 2003. At the national level privatisation has also been reversed in Uruguay, for example (Hall *et al.* 2005, 292), and in the Netherlands water privatisation was even made illegal in 2004 (Hall *et al.* 2004).

In fact, despite the waves of privatisation from the 1980s onwards, overall private sector participation remains limited and the prospects for growth in the future poor (Swyngedouw 2009b, 51). Indeed, water multinationals such as Veolia and French rivals Suez have been gradually withdrawing from the market since 2003 (Hall and Lobina 2008, 3). Despite this trend and moves to re-municipalisation, the future of privatisation in WSS is not clear-cut. It is still being implemented in countries such as China and in Spain it has steadily grown over the past few decades (Sauri *et al.* 2007). In Italy the government introduced legalisation in November 2009 that would open up the publicly owned sector to private investment (Reuters 2009). This initiative was, however rejected by the public in a referendum in 2011.

Privatisation continues to be implemented around the world despite the experiences of England and Wales suggesting that privatisation of WSS is inherently problematic. After two decades of privatisation the objective of creating market conditions in WSS is still far from being achieved. Initially imagined as a sector only lightly regulated, a process of re-regulation occurred throughout the 1990s, with the economic regulator, the Water Services Regulation Authority (Ofwat), taking an increasingly interventionist approach. This was provoked by unease at one of the necessary consequences of privatising WSS: the profits made by the private water companies. Public and political unease grew as water companies registered high profits while prices increased and the affordability of water for low-income customers became an issue (Bakker 2003b, 8).

The development of competition has also been stunted. Given the natural geographic monopolies enjoyed by water companies and the technical and environmental obstacles to introducing competition for supply and sanitation within these monopolies, private sector companies have limited options for growth, particularly if the regulator rules against price increases. At times some water companies have struggled to finance investments imposed by Ofwat to meet water quality and environmental standards whilst returning profits to their shareholders (Bakker 2003b, 15). As well as prompting water companies to seek ways of reducing costs – outsourcing services and cutting jobs – three key strategies of restructuring emerged: diversification, internationalisation and, most radically, returning ownership of assets to not-for-profit companies, while retaining responsibility for water supply and treatment. Ofwat received two proposals of this kind: from Kelda Water (formerly Yorkshire Water), which was rejected and *Dwr Cymru* (formerly Welsh Water), which was successful.

Alongside these moves to – at least partially – reverse privatisation, the sector rests on a more fundamental tension that has arisen as a result of private ownership: between the environment as a "legitimate user whose interests are to be balanced with consumers" (Bakker 2003b, 27) and the companies and their shareholders searching for profits. This tension centres on the "argument about trade-offs between cost and environmental improvement: *How much improvement? At what cost? Who pays?*" (Maloney 2001, 633). As Bakker (2003a, 2003b) has argued, the privatised water regime has failed to address key contradictions between stable returns for water companies (and their shareholders) and the efficiency imperative (reduce consumption and leakage), politically acceptable profits and issues of equity, the mounting requirements of environmental regulation and the profits of shareholders as well as the prices of water. As a response to this there has been broad restructuring of the sector – diversification, internalisation, vertical de-integration, mutualisation and securitisation. Further debates have emerged around the desire to introduce greater competition into the sector.

Conclusion: re-politicising privatisation

Given the highly mixed experiences of privatisation in the WSS sector, it cannot be viewed simply as a pragmatic, commonsensical policy. Furthermore, privatisation in the WSS sector has resulted in corruption. For example, Veolia, one of the largest water companies in the world, has allegedly been involved in a number of cases of bribery of public officials since the 1990s: for example, Strasbourg (1991) and Saint Denis (1996) in France and Milan in 2001 (Public Citizen 2005). It is has been acknowledged even by the World Bank, one of the prime supporters of privatisation in the water sector, that privatisation projects often create conditions in which corruption can flourish, with the prospect of lucrative contracts and sale revenues encouraging bribery (Hall 1999, 11-12). Nonetheless, this has not stopped some researchers advocating privatisation as a solution to corruption (Rose-Ackerman 1999).

How can the continued promotion and implementation of privatisation be explained, given the potential for corruption and failure to achieve stated objectives? Privatisation of WSS services is not a policy which usually enjoys widespread public support. It has been strongly contested, rejected and reversed around the world since its rise in the early 1990s. It has the capacity to divide, with contradictory empirical evidence produced by opponents and supporters. If privatisation cannot be objectively proven to be superior to public forms of management, then it must be understood in terms of the theoretical and normative claims that underpin it. As Brown (2005, 38) puts it, apparently economic policies such as privatisation are organised by the broader political rationality of neo-liberalism. This rationality describes the world in terms of market rules and actively seeks to realise such a world. It is fundamentally *political*. Privatisation and other policies rest upon normative not ontological claims about the world.

The increasing prevalence of privatisation should be explained with reference to the potency of neo-liberalism. Privatisation may often be, as Barraque (2009, 245) states, opportunistic, in that governments may not be obviously driven by 'ideological' arguments, but simply want to raise money or sell a loss-making company. Obviously, the private sector also has an interest in promoting privatisation given the financial opportunities it provides. It also has the financial means to lobby for privatisation and exert pressure for privatisation. It is still, however, necessary to justify policies in representative democracies and this might often entail the provision of broader, moral, theoretical or evidence-based arguments. The political context of the 1990s made it increasingly easy for governments to present such arguments about privatisation because neo-liberalism was promoted and accepted by many as commonsensical (Peck and Tickell 2002, 34).

As stated, this thesis does not attempt to explain the pervasiveness of neo-liberalism in this period. Rather, the aim is to analyse one privatisation within this broader context. By focusing the analysis on the policy-making process, a key task for the thesis is to expose the effects of neo-liberalism: the way in which normative claims shaped the making of facts and a politics of inevitability. It is concerned, more specifically, with the presentation of 'facts' in policy-making processes. In the BWB case, it is necessary to explore the ways in which privatisation was presented by its supporters: how 'facts' about BWB, the wider context in Berlin and the benefits privatisation would bring were constructed; how opposing arguments were countered, their promoters sidelined in the policy-making process and enough support secured for privatisation to be implemented.

Chapter 3. Facts and values in policy-making

Introduction

Despite the controversy one could expect with a partial privatisation of water supply and sewage services, particularly in a 'left-ish' city like Berlin, the BWB privatisation was presented by the government as commonsensical and its implementation was, according to interviewees, relatively straightforward. Given the debates over the effects of privatisation and the controversy it has provoked, the presentation of privatisation as a panacea by governments is misleading. It is, then, necessary to move the analysis of the BWB partial privatisation beyond the claims made by policy-makers and focus on the actualities of making policies. Policies such as privatisation are not neutral and cannot be objectively 'good'. They have to be constructed as such. A full understanding of policy-making and a shift to privatisation can only be achieved through an appreciation of the way in which normative and theoretical assumptions inform the making of facts in policy processes.

This brings the discussion to broader issues researchers have raised about contemporary representative democracies. In particular, it is necessary to address the claims to 'rationalism' and 'objectivism' that underpin the formal institutional policy-making processes (Dryzek 1994): to move away from notions of policy-makers and political systems as entirely rational and objective. A concern with the workings of policy-making also necessitates a consideration of contemporary discussions about 'governance': the perception that not only institutions of the state participate in governing but that business and societal actors can and should play a role in a more fragmented system of governance. Moreover, it is necessary to move beyond depoliticised notions of governance: research must focus on the actualities of governance systems; to reveal the effects of neo-liberalism in shaping them.

The task is to conceive of policy-making in an "institutional void", where there are no "generally accepted rules and norms according to which policy-making and politics is to be conducted" (Hajer 2003b, 175). Even if this is, as Rhodes (2006, 438) states, an overly dramatic assessment of governance in contemporary representative democracies, it underlines this sense that roles and responsibilities in policy-making are not only derived from the formal institutions of government. Consequently, the researcher should pause before making too many assumptions about how policy is made, by whom and in what fashion. Rather, to understand contemporary democracies it is necessary to trace patterns of governing and reveal the relative status of public and private actors.

This chapter centres on a discussion of facts and values in both the making and the analysis of policy. It begins by problematising rationalism and objectivism. It then argues against the positivism that characterises much of the policy studies literature, proposes the adoption of a constructivist theory of knowledge and an approach focused on the construction of facts and truths in policy-making and governance. This is done with reference to the interpretative policy analysis literature, particularly the work of Maarten Hajer, Frank Fischer and Dvora Yanow. 'Facts' are shown to be influenced by normative assumptions and strategies of action. Experts are highlighted as being central to the production of facts in politics. With the rise of neo-liberalism, and more particularly 'managerialism' in government, private sector consultants have become more influential in policy-making.

The relationships between knowledge, values and policy-making

The contestation of existing political practices and the reforms that sometimes emerge from them is a defining feature of governing in democracies. Governmental practices and state-society relations are defined through contests between competing claims about the world and the role of government within it. With such claims, a government's ability to convince the electorate that a publicly owned company is inefficient or a bank's ability to convince the government that intervention in the financial sector impedes performance, come proposals for reform. Policies, such as privatisation or de-regulation of the economy, emerge as remedies for problems, but they also make claims: claims about both the current and future state of things. Simple examples might include: *The publicly run company is inefficient, badly managed – with privatisation, the company will be more profitable, provide better services; Red-tape increases transaction costs – with de-regulation the economy will be more flexible, profitable and dynamic.* Policy-making is about defining an aspect of society, but it is also a forecasting activity, outlining a means to reform it, with predictions as to how that world will be. Policy-making therefore involves proposing a new, improved version of the world along with the means to achieve it. It brings together knowledge claims about the present and theories about how the future will be.

Rationalism and objectivity in policy-making and policy analysis

A challenge for any analysis of government is to explain how such claims are formed (where they come from) and why some claims resonate more than others at particular times (why they become policies). Underlying this, of course, is a concern for how political power is exercised, how it becomes wrapped up in these claims about the world and how policies and new practices of government emerge from this interplay.

At first glance there appear to be two distinct concerns in an analysis of government. First, how knowledge is produced: the ways claims about state and society relations are made and the ways in which they become accepted by others, either as descriptions of reality or as proposals for the future. Second, how political power is practiced: how these claims become accepted; how proposals become policies; how governmental practices change.

One possible way to start an analysis would be by suggesting that some claims about the world are more truthful than others, more rooted in the hard facts, and this is why they tend to 'stick'. For example, by the time of the official privatisation debate in 1997-1999, few in Berlin contested the need for the government to raise cash to reduce the ballooning debt. Many accepted the view that BWB had, for a few years and at least in certain respects, been poorly managed. With these claims about Berlin and BWB a series of problems had been identified and agreed upon. Given these problems and the widespread agreement upon them, privatisation (as a means of raising cash and improving efficiency), simply became seen as the logical solution. Or so it is often suggested (by Fitch 2007a, in the case of BWB). There is, of course, a process missing from such an explanation: the step which explains why privatisation became the proposed means to solve this problem and more subtly, why, and in what ways, BWB's status and performance came to be linked to Berlin's finances and future. The point here is that the links between the two had to be made. In the process of making these links alternatives to privatisation were marginalised. The privatisation proposal was merely one representation that relied on certain assumptions about the status of public companies, the role of government and the benefits of the private sector.

Privatisation, or any policy, emerges, in part, from some kind of theorisation about the correct, appropriate role of government. However much it is presented as a pragmatic policy tool, it is also rooted in intellectual and normative claims. Following this line of thinking, if claims about the world emerge in relation to the facts around us (as a direct representation of reality), what to do with this reality – the proposals to reform it – emerge from intellectual and moral deliberations. There is then an apparent separation between the gathering of facts and the intellectual and moral means of analysing those facts to make policy proposals.

From such a perspective, this gap, the critical distance between facts and interpretation, is seen to allow for an objective assessment of facts and the potential for rational decision-making. Thus, though proposals may rely on intellectual and normative claims, facts exist independently. Accordingly, policy proposals must be determined and assessed in terms of their relations to the facts – how well they correspond to reality. In other words, because facts can be objectively apprehended, policy proposals themselves can aim for rationality and objectivism. Indeed, this is the fundamental assumption made by much of the policy-making community and those who study policy-making (see, for example, Sabatier 1999). According to Dryzek claims to rationality can be seen as underpinning both the practice and study of politics: "individual political behaviour, the policies of governments, the structure of political systems, and the assumptions and strategies of political scientists who study these phenomena are all more or less rational" (Dryzek 1994, 3).

Thinking first in terms of policy-makers, Dryzek has identified two defining and problematic manifestations of rationality in policy-making in liberal democracies: "instrumental rationality" (the 'rational' selection of means to achieve specific ends) and "objectivism" (Dryzek 1994, 3-4). The former refers to the way political systems are organised while the latter refers to the logic of decision-making, which is seen to guide this system. As Chapter 6 will show, both features are apparent in official accounts of the BWB privatisation and more broadly in the 'no alternative' to privatisation argument put forward by many governments in recent years. Underlying these justifications is an elitist belief that the political system knows better and a sense that there is, despite any contestation that might arise, an objective political or even moral standard by which policy-making can be measured (Dryzek 1994, 4).

Such a faith in, and quest for, objectivity is also apparent in the approaches employed by researchers of policy-making. As Danziger (1995) and Fischer (1998) state, much of the work in this field has utilised a positivist theory of knowledge (e.g. Doron 1992; Klingemann and Hofferbert 1994; MacRae and Whittington 1997; Lijphart 1999; Lynn 1999; Ostrom 1997; Putt and Springer 1989; Sabatier 1999; Sabatier and Jenkins 1993; Weimer 1999). Positivism is based on the assumption that there is a reality independent of human interference, one constituted by general laws of cause and effect that can be exposed through the empirical testing of hypotheses (e.g. King *et al.* 1994). Such studies often claim to be objective and value-free as the general laws they seek to expose exist independently of social and political contexts (Fischer 1998, 28). The focus is largely on the role of interest groups, political leaders, bureaucrats, agendas and decisions (Wilson 2000, 248).

One of the most popular and influential positivist approaches to policy analysis is the Advocacy Coalition Framework (ACF), developed by Sabatier (1999) and Sabatier and Jenkins (1993; 1999).[13] It has been applied widely, most particularly to the study of environmental and energy policy, education, and national defence (Sabatier and Jenkins-Smith 1999, 125). It can be seen as an attempt to advance empiricist studies within the field, moving beyond the conventions of the 'stages model' (agenda-setting, policy formulation, adoption, implementation and evaluation) of policy studies (Fischer 2003, 95).

Sabatier and Jenkins-Smith aim to provide a causal model that can penetrate the complexity of the policy process, explaining policy change through a behaviourist assessment of the interactions between competing 'advocacy coalitions' and the impact of external events upon these coalitions. These advocacy coalitions consist of actors from a range of institutions who share a set of policy beliefs and engage in some form of activity together within a policy field or 'subsystem' (Sabatier and Jenkins-Smith 1999, 9). The aim of the framework is to allow the policy researcher to first ascertain the "belief systems" of actors, organise them accordingly in coalitions and then analyze the conditions under which they can morph, facilitating policy learning across coalitions (Sabatier 1999, 9).

Sabatier (1999, 4) seeks to distinguish his own model from Institutional Rational Choice approaches (such as Scharpf 1997; Shesple 1989), which he argues offers too narrow an explanation of political behaviour and too limited a scope in terms of actors relevant to the analysis. These approaches focus on leading figures within the key institutions holding formal decision-making authority, assume that these actors are motivated only by self-interest (income, power and security) and tend to group them according to their institutional affiliations such as legislatures, administrative departments and interest groups (Sabatier 1999, 4). The assumption here is that institutional affiliation goes a long way to explaining the identities of actors, providing the basis for their interests. In contrast, the ACF assumes that 'belief systems' are more important to actors' identities than institutional affiliation. As a result they may pursue a wide variety of objectives that can and should be measured empirically and that a broader set of actors, namely policy analysts, researchers and journalists, are potentially important to policy-making given their roles in shaping belief systems (Sabatier 1999, 5). To capture the alliances between public and private sector actors there is therefore a need to move beyond conventional notions of policy-making such as the 'iron triangle' of administrative agencies, legislative committees, and interest groups operating at a single level of government (Sabatier and Jenkins-Smith 1999, 119).

Sabatier and Jenkins-Smith do attempt to tackle the issue of knowledge production, albeit from a positivist perspective. They argue there is a need to focus upon knowledge – "technical" (expert, scientific and policy) information – to

[13] Other key approaches informed by positivism include Policy Diffusion, Institutional Rational Choice, Multiple Streams, Punctuated Equilibrium and Large N-Comparative Studies (see the edited volume by Sabatier 1999 for prominent examples).

expose the crucial roles it plays in defining the 'truth' in debates (Sabatier and Jenkins-Smith 1999, 118). In terms of analysing the actual content of policy proposals – and the contests between them – they argue that public policies embody theories about how to achieve objectives and therefore can be conceptualised in much the same way as belief systems: they involve value priorities, perceptions of important causal relationships, perceptions of world states" (Sabatier and Jenkins-Smith 1999, 119).

In such positivist perspectives, policy-makers react to the facts around them. Reality shapes their beliefs and they then act – seek to make policy – according to their beliefs. This is where political philosophies, belief systems or ideologies such as neo-liberalism normally come into the analysis. Political parties do, at least nominally, adhere to political philosophies or ideologies, however much and in whatever ways these may change. Adhering to, or advocating, these different philosophies explains why there are differences in policy proposals. Why the Left might advocate more state intervention in the economy, and why the Right might counter that with an argument for the profitability of de-regulated or self-regulating markets. Political parties have overall objectives, a set of values and a vision of society that they present to the public.

The changes of parties' political programmes and shifts in ideology are explained with reference to either a lack of success with the electorate or in more theoretical terms, the diminishing resonance of that philosophy with reality. In the ACF perspective socio-economic conditions rather than institutions or strategic interactions are seen as a catalyst for change. Broader changes in societies transform public opinion at large, which, in turn, affects the policy preferences of interest groups, political parties and ultimately the mindsets of decision-makers (Fischer 2003, 97).

Actual policy change then is conceptualised as occurring through 'shocks' that disturb settled beliefs systems and relatively stable political behaviour. In such instances, the new problems confronting actors seem not to fit existing solutions, controversies emerge and advocacy coalitions lose their solidity. Such periods when core beliefs are challenged are seldom (the oil crisis of the 1970s is given as an example) and in such cases Sabatier argues that it is difficult to predict policy outcomes. Actors generally want to hold consistent beliefs, which explains why changes in core beliefs tend to occur only through these external shocks.

Another influential approach, Baumgartner and Jones's (1993) Punctuated Equilibrium Theory, also views policy change as emerging from "external shocks". Such events shatter the normal stability of the political system, disrupt the standard pattern of incremental policy change and bring about more radical reform. In these periods existing policies are contested and the belief-systems that have underpinned them can be changed. The rise of neo-liberalism and the decline of the Left in many countries have often been explained in such terms. Similarly, the emergence of the

Third Way has been seen as a strategic and philosophical shift to make the Left more 'relevant' and popular. Implicit to these assertions is the separation of 'reality' from the means of analysing it and the development of proposals to govern it.

As Larner (2000) states, this is the most common form of conceptualising neo-liberalism: as a "policy reform programme initiated and rationalised through a relatively coherent theoretical and ideological framework" (7). This policy framework entails the shift from Keynesian government to a form of government in which the market – to varying degrees – is seen as the best way of organising state-society relations in the context of globalisation (Larner 2000, 6). As discussed, the key argument put forward is that globalisation – particularly of finance – has put governments under pressure: forced them to re-consider their roles in the societies they govern. The above manifestations of neo-liberalism are the response to this 'reality' of globalisation. The pre-dominance of neo-liberal forms of government is explained largely with reference to the ascendancy of political and economic actors adhering to a particular ideology or worldview promoting the individual, freedom of choice and a restricted role for government to best facilitate these objectives (Larner 2000, 7). Accordingly, the rise of these ideas is also primarily explained with reference to the influence of key politicians (e.g. Thatcher and Reagan) and organisations (e.g. the International Monetary Fund {IMF} and the World Bank), with the support of intellectuals and the financial muscle of multi-national capital.

Drawing these distinctions between facts, values or ideologies and power is certainly a useful means of breaking down a complex process into something more easily analysed and understood. As such it is understandable that policy-makers and implementers as well as policy analysts aspire to this objective approach. But is it really a convincing account of the way these things relate to each another in the making of policy?

Larner (2000) criticises this mainstream perspective for being too focused on politicians and policy-makers; too focused on the outcomes of neo-liberal policy programmes (the 'were they successful or unsuccessful?' debate); and therefore lacking the capacity to explain the broader phenomenon of neo-liberalism. Such approaches reveal little about the role of ideas in neo-liberalism. Furthermore, they have a tendency to underestimate the effects of neo-liberalism in actually shaping reality – in shaping the production of facts. For example, in normalising "the logics of individualism and entrepreneurialism, equating individual freedom with self-interested choices, making individuals responsible for their own well-being, and redefining them as consumers and clients" (Leitner *et al.* 2007, p 2).

Mixing facts with values: producing realities, making policies

If a closer look is taken at the interplay between facts and values in policy-making, any separation between the two becomes problematic. What comes first in a policy-making process, the claim about how the world is (the production of knowledge) or the values and theories of politics (how that world should be re-created)? Often in policy debates they seem to come together, feeding off one another, one explaining the other's existence: the debts of Berlin leading directly to the privatisation of public companies, the apparent superiority of private sector management, highlighting the inadequacies of BWB's public management. Thus values shape the making of problems and linking of solutions to those problems. Political philosophies, norms, values and so on shape not just the policy proposal but also the claim about reality on which it is based. Values play a role in the making of 'facts'; theories about the world shape perceived realities.

Consider an alternative policy proposal to that of the partial privatisation of BWB. As will be shown, in 1998 the Unions made a proposal to retain the public status of the company, but to merge it with the Gas Company (GASAG). This proposal accepted that BWB should be reformed and that money should be raised from this reform to ease Berlin's debts. The plan was that a multi-utility company would be more efficient in terms of manpower costs and more effective in bidding for work. So, it accepted aspects of the claims made by the supporters of privatisation. However, it did not accept that the company should be sold, that private ownership and management would be more efficient. It did not accept the 'logical' linking of BWB's and Berlin's problems with the private sector solution.

It could be said that competing policy proposals see the same things differently – that they see the same world but, drawing on different values and theories, come to different conclusions about how to reform it. More accurately, however, it should be argued that, as a result of these moral values and theoretical tools, competing policy proposals see reality differently: a reality composed of different elements, working according to different logics and norms. A reality in which public management of BWB may have been deficient, but not all public management is necessarily deficient. Thus a competing policy proposal, as well as proposing a different solution, may also represent the problem in a slightly different way; make a slightly different claim about the world based in this case on an alternative normative view of private sector involvement. The differences between these claims must be said to lie in the different values and intellectual tools used to render them real.

Accordingly, the task for any analysis is to examine not how facts inform the making of policy proposals, but how facts about an aspect of the world are made; how knowledge is produced and how, in turn, this knowledge shapes policy options and ultimately governmental practices. Policy-making is as much a process of making a world real as it is governing that world. Or, to put it in a slightly different way, a world has to be represented before it can be governed. Policy-making is a representation of an aspect of the world that simultaneously proposes the practical means to reform that world, to create it anew. As such, an analysis of policy-making requires an assessment of the complex relationships between knowledge, values, theories and power. It is concerned with how the production of knowledge is interwoven with the expression of values, the articulation of discourses such as neo-liberalism and the exercise of power. The following section lays the foundations of an approach to tackle these issues. It takes on the issue of facts and values in policy-making, outlining a constructivist theory of knowledge and demonstrating how this shapes a conceptualisation of policy-making and government.

Moving towards a post-positivist approach to policy studies

In recent years positivist policy analysis has been increasingly contested by what has been variously called post-positivist, post-empiricist, critical or interpretive policy studies. This literature can refer to hermeneutics, phenomenology, social construction, critical theory, discourse analysis and deconstruction (Yanow 1995, 111). Key writers include Yanow (1995; 2003), Fischer (1998; 2003; 2009), Hajer (1997; 2002; 2003a; 2003b), Howarth (2005), Torfing (2005), Gottweiss (1999; 2003) and Dryzek (1994). The growing influence of this literature is such that prominent political scientist Rod Rhodes has sought to re-align his work as 'interpretivist' (Bevir and Rhodes 2006) (a move both welcomed and critiqued by Glynos and Howarth 2008).

Though various in the approaches it encompasses, this body of work can be said to begin with a challenge to the positivist presupposition that simple observation can provide "ready access to an objective world" (Yanow 1995, 111). More broadly, there is a rejection of the modernist stress on rationality and objective knowledge. "Facts, in the natural as well as the social world, depend upon underlying assumptions and meanings" (Fischer 1998, 11). This does not mean that there are not aspects of reality, which are separate, independent from researchers, but rather that the "vocabularies and concepts used to know and represent them are socially constructed by humans beings" (Fischer 1998, 14).

Meaning in policy-making is not taken for granted. Indeed, the ways in which meaning is created, communicated and understood become a focus of study. The question for policy studies is, then, to follow the title of one of Yanow's (2003) books, *How does a policy mean?* How do we apprehend the meaning of policies and how do policies provide meanings? The starting point for such an enquiry is that policy is a social construct: that people, through a variety of practices, create their own social realties. Furthermore, researchers far from being able to stand back and coolly analyse them, can only offer contingent interpretations of those realities. In this sense, objective or universal knowledge about the world is replaced by an appreciation that the production of knowledge is always contingent upon actors' own interpretations.

As Bevir (2006; see also Bevir and Rhodes 2006) states, seeing policy as a construct has the effect of "de-centring" governance. It shifts the analysis away from institutions and the norms that are embedded in them, to an examination of the ways in which individuals "create, sustain and modify social life, institutions, and policies" (Bevir 2006, 24). Change in policy is seen as the result of contests over meaning: the reality of things and the relating of policies to those realities. Contests over meaning are so integral because different theories, political rationalities and normative judgments produce both different realities of government (where some see faults others see success) and different proposals for reform. A key aim of politics is the transformation of an existing reality and, as Fischer (1998) states, much of the "struggle turns on the socio-political determination of the assumptions which define it" (12). Governance is ongoing, defined by these changes because most policy will eventually become contested (Bevir 2006, 25).

Viewing policy in this way shifts the focus of the analysis onto the creation of these meanings – onto the production of truths in politics. Reality is no longer, simply 'out there', waiting to be apprehended and reformed by policy-makers. From such a perspective Sabatier's ACF, despite its attempts to move beyond rational actor approaches, is still hindered by its positivist theory of knowledge and quest to create a causal model of change. Studies from this perspective may have revealed that interests alone do not explain change (Fischer 2003, 19) but they can, however, be criticised on their own positivist terms for not offering a convincing account of why policy actually changes (Fischer 2003, 99). Crucially, change is mainly initiated by 'events' but there is little discussion about how these events are interpreted, distorted or even created by actors. Instead, they come across as simple facts that actors either do or do not react to. There is no sense in which the importance of these facts must be constructed by policy-makers to initiate change in policy-making. This may explain why empirical studies utilising ACF have often revealed that 'shocks' have not always led to the expected disruption of coalitions (Fischer 2003, 99).

What is absent from the ACF is a focus on the construction of facts and a sense that language plays a role in the shaping of facts. There is no such thing as theory-neutral observation language, no simple reality that can be represented neutrally (Gottweiss 1999, 63). Instead "the 'truth' of an event will always be the uncertain outcome of a struggle between competing language games or discourses which transform 'what is out there' into a socially and politically relevant signified" (Gottweiss 1999, 63). For this reason an appreciation that discourse is constitutive of reality is essential to the analysis of policy. Adopting a discourse perspective transforms the substance of policy. Light is now cast on language (texts, speeches and policy documents) not merely as an expression of discourse hegemony but as constructive of discourse hegemony. As Hajer (2003a, 103) states all discourse analysis "aims to show how language shapes reality".

Overall, then, what is missing from the ACF is a convincing account of the ways in which actors try to make sense of the world around them – the social and historical context of policy-making. In its "empiricist's desire to develop empirical hypotheses that are universally applicable to the widest range of social factors" (Fischer 2003, 101), the ACF ignores the "symbolic, normative concerns that a discourse analysis takes as its primary subject" (Fischer 2003, 101). Indeed, Sabatier has explicitly rejected such concerns on the grounds that they are not theoretically robust enough (Sabatier 2000).

The aim of a post-positivist analysis of policy change is to move beyond a mere "fault-finding" assessment, characteristic of much of the mainstream, positivist literature, to that of critique (Bevir 2006, 28). Critique, drawing on in-depth interviews or detailed documentary analysis, aims to get under or beyond official accounts – as Chapters 5 and 6 will – and provide "texture, depth and nuance" (Bevir 2006, 28). From such a perspective, privatisation is never simply a pragmatic response to circumstances but a response that makes claims to being pragmatic that conceal neo-liberal normative and theoretical biases. The analysis of the BWB partial privatisation will then be less an evaluation and more an unmasking of policy-making as "contingent, partial, or both" (Bevir 2006, 22).

This form of analysis can act as an important counter balance to positivist studies that overlook the processes through which particular realities are constructed – the theoretical and normative assumptions underpinning them. Such studies risk reifying a particular reality and, "wittingly or unwittingly", could serve as much to provide ideological support for a policy as to explain it (Fischer 1998, 12). A post-positivist, interpretivist approach is necessary to get beyond the claims to rationality and objectivism that policy-makers tend to make.

The rationalism and objectivity of policy-makers and political systems

In the case of the Berlin government of the 1990s we should be particularly sceptical of such claims, given its record of financial mismanagement and corruption. As such, the official account of the BWB privatisation leads into bigger questions about the organisation of politics and policy-making in contemporary liberal democracies. Of course, criticism of bureaucracy and government is nothing new and can be traced back most notably to Weber and his fear of the rationalisation of politics and society in capitalist countries (Dryzek 1994, 4-5).

Weber warned that the assembling of administrative resources to manage societal problems would lead to a concentration of political power. Integral to this tendency towards centralisation in governing has been the emergence of technical experts in an ever-growing range of specialisations: the rise of administrators, marketing consultants, lawyers, social workers as 'experts' in policy-making (Dryzek 1994, 5). This combination of bureaucratisation and utilisation of technical expertise has led to criticism of the 'rational', 'technocratic' governance of liberal-democracies; a form of governance that is inherently "anti-democratic" (Dryzek 1994, 7).

More recently, interpretivist policy analysts and public administration researchers have located the perceived failings of representative democracies within broader critiques of 'modernism' (see, for example, Dryzek 1994, Fischer 2009, Hajer 2003b, Miller and Fox 2007). Such work points to the policy-making deficiencies and the elitist tendencies of the logics and institutions of modernism. This has often led to discussions of the capacity of modernist logics and institutions to govern contemporary society, be that one understood in the terms of Beck's "risk society" (Hajer and Kesselring 1999) or Castells' "network society" (Hajer and Wagenaar 2003). Such arguments rest on the view that in an intrinsically dynamic society, experiencing, amongst other things, globalisation, the legitimacy and effectiveness of traditional political institutions is being increasingly challenged (Hajer 2003b, 177).

'Classical-modernist' political systems

As meanings are always contingent, dependent upon the interpretative work of actors in particular contexts, it is argued that there can never be truly rational, objective decision-making. To make such claims endows political systems with a false sense of certainty and legitimacy. Dryzek (1994) has potently argued that these two strands of thinking cast a shroud over the realities of public policy-making, concealing in-built biases, the privileging of certain knowledges over others and more generally the inability of modernist political institutions to deal with complex problems.

The core of the "classical-modernist" democratic system is "representative democracy, a differentiation between politics and bureaucracy, the commitment to ministerial responsibility and the idea that policy-making should be based on expert knowledge" (Hajer 2003b, 176). The system rests on the "representative democratic accountability feedback loop" whereby individuals are seen as being capable of forming rational preferences, which are then "aggregated to popular will, codified by legislation, implemented by bureaucracy and evaluated in turn by attentive voters!" (Miller and Fox 2007, 122). Within this system, decision-making is dependent on a "a stable but small set of actors when it comes to preparing policy plans combined with the assumption of a smooth implementation of any well-considered plans" (Hajer 2003a, 92).

Adopting a discourse analysis approach, Hajer emphasises that the political institutions of contemporary representative democracies should be seen as historically contingent – as a product of the post-war modernist era and not as an immutable feature of democracy per se (2003b, 176). This distinction is important as the forms representative democracies have taken are not seen as being absolutely fundamental to democracy itself; instead, they are an articulation of democratic decision-making in a particular place within a particular historical period. Adopting such a view avoids the reification of existing arrangements and focuses the researcher on the contextual conditions through which they are produced and re-produced. Of course, this is only an image of policy-making and the modernist thinking which informs it. Indeed, Dryzek (1994, 33ff) has argued that "critical rationalism" better represents the current thinking informing policy-making and administration in many representative democracies. Still, as is revealed in Chapter 6, such images of policy-making remain potent to those who make policy.

Reforming the classical-modernist system

As stated, misgivings about the classical modernist system have long been aired but the conclusion has often been to say that the model itself is fine; that the problems lie in the people populating and abusing it (Miller and Fox 2007, 3-4). It has therefore been argued that reforms should aim to "resuscitate it by making it still more efficient, rational, scientific and/ or progressive" (Miller and Fox 2007, 4). The importing of private sector practices and personnel in shifts to "managerialism", "market-based administration", the "entrepreneurial government" or the New Public Management (NPM) (see Barzelay 2001), from the 1980s onwards can be interpreted as one such move. NPM aims for efficiency reforms through organisational restructuring; management instruments; budgetary reforms; participation; customer orientation and quality management; and marketisation and privatisation (Schedler and Proeller 2002, 165).

Saint-Martin (2000) stresses that this "new managerialism" is emblematic of the "political change to governance" (Saint-Martin 2000, 1). In other words, moves to governance should not be seen in neutral terms, as mere responses to a 'need' for greater efficiency and expertise in government. Saint-Martin (2000, 198) goes on to emphasise that there is no "politics-administration dichotomy": no truly neutral form of decision-making and bureaucracy in political systems. Managerialism may not have been the invention of governments led by Reagan and Thatcher (Saint-Martin 2000 reveals forms of managerialism emerged earlier and sometimes under left-of-centre governments) but as Miller and Fox (2007, 30) state it certainly "picked up steam" as these governments implemented neo-liberal policies. Moves to managerialism in political systems have thus been informed by neo-liberalism. They have, in turn, been crucial to the processes through which neo-liberalism has been realised.

Consultants and governance

One discernable feature of these shifts in governing, from the centralised, hierarchical Weberian bureaucratic model to an apparently more flexible, market-based form of public management, has been the emergence of private sector consultants as a prominent form of expert in policy-making and administration (Saint-Martin 2000, 1). Along with the outsourcing of services, privatisation and the development of public-private partnerships, the use of consultants is one of the main methods by which the private sector has entered into new governance structures (Cook 2009, 935-6). As stated in Chapter 1, they played prominent roles in the BWB partial privatisation. They were lobbyists for public and private actors. They were also employed at great expense by the government to manage aspects of policy implementation. They are then integral to the analysis. The study therefore explores their roles in the BWB privatisation process, linking them to the overall study of neo-liberalism and shifts to governance.

Despite their increasing importance in politics, the literature on consultants is relatively limited. In general terms the increase in consultants in government has been attributed first to their increased prominence in the private sector – the apparent need for greater professionalism and expertise in business (Hodge and Bowman 2006, 97). The spread of consultants from business into government itself has been seen as the result of a two-sided, supply and demand process (Moran 2006, 157). The supply has been created by the fierce competition between (particularly multi-national accountancy) firms in the service sector; on the demand side, governments have increasingly sought to introduce 'businesslike' practices to government (Moran 2006, 157-8). This shift to a NPM or, more generally, a managerialist approach to government, has been underpinned by a pursuit of greater efficiency and a belief that consultants provide specialist knowledge not found within the public sector. The privatisation movement can be said to embody both the supply and demand sides of this process. It has been the key tool through which governments have sought to reform their roles, while the "marketing of expertise in the privatization process has been an important means through which the phenomenon of privatization itself has been diffused" (Moran 2006, 157).

Governments are increasingly employing consultants to perform a huge variety of tasks. As will be shown in 1990s Berlin, consultants played important roles across a range of policy-making sectors including privatisation, water management, city planning, banking, urban development and marketing. More generally, in the European context the rise of consultants in government is most evident in the UK (European Federation of Management Consultancies Associations 2006). Defining consultants narrowly in terms of those individuals or companies employed to work on specific projects, a recent report from the UK National Audit Office (NAO) stated that the public sector spent around £3 Billion ($5.8 Billion) on consultants in the financial year of 2004-2005. This represented a growth of around a third in two years, with consultants working in every sector of public administration. More recently a survey of consultants working in the environmental sector revealed that, for the first time, most firms expected the public sector to provide more opportunities for new work than the private sector (Edie 2009).

In Germany, public administration has been characterised as being generally more resistant to managerialist trends due to its well-developed, legally defined framework for governing (Peters and Savoie 1998, 8). NPM reform, as well as being minimal in West Germany, occurred quite late in comparison to the UK and Netherlands, emerging in the late 1980s and early 1990s and more at the local level than that of the *Länder* (States) or Federal government (Schedler and Proeller 2002, 172). However, consultancy services for the German public sector are now one of the biggest markets of its kind in Europe. The Federal Association of German Consultants (BDU) noted that in 2005 the German public sector spent in the region of €1.28 Billion on consultancy services (website).[14] Between 2007-2008 public sector bodies were the main clients for consulting services (with 38% of the total market) in Germany (European Federation of Management Consultancies Associations 2007, 19). The public sector was also the main market for the UK, Denmark, Greece and Hungary (European Federation of Management Consultancies Associations 2007, 19). As Hodge and Bowman (2006) have stated, there is now a "global business in reforming government" (97) with consultancy firms playing an increasingly prominent role in "providing the intellectual backbone of government reform packages" (98).

The limited academic literature on consultants and politics has raised concerns about their roles in political decision-making. Thus the rise of consultants is often seen – in critical tones – as having led to the emergence of a "consultocracy" (Hood and Jackson 1991, 24 in Saint-Martin 2000, 20): the commercialisation of government and increasing use of the business sector as advisers in policy-making. However, criticism of consultants in government is not an altogether new phenomenon. In the 1970s, Guttman and Willner (1976) criticised what they saw as the emergence of a "Shadow Government" of

[14]BDU: Bundesverband Deutscher Unternehmensberater e.V.

management consultants in American politics. This, they argued, raised questions about conflicts of interest, use of public money to pay for such services and the development of a secretive decision-making apparatus within government.

More recent criticisms are of a similar nature, stressing that the traditional processes of making policy have become "obsolete", with consultants selling policy solutions, "underneath a less democratic policy development process (Hodge and Bowman 2006, 109). Their presence in government, within new forms of governance (often partnerships), has led to criticisms of private control over public policy (Shaoul *et al.* 2007). Furthermore, this increasing use of consultants has resulted in the growing use of private sector practices in politics. Most controversially, the use of commercial 'in confidence' agreements between public and private sector (of which the BWB partial privatisation deal is a good example), as well as more confidential meetings involving consultants and key politicians, both of which cordon-off decision-making from Parliament and the public (Hodge and Bowman 2006, 108). Such secrecy and lack of transparency raises issues regarding the accountability of policy-making. Ultimately, the rise in consultants is seen to "parallel a decline in democracy for the citizenry" (Hodge and Bowman 2006, 108). Their increasing importance in policy-making, their presence within, or alongside political institutions, is seen to have undermined the capacity of political systems to ensure transparency and accountability in policy-making.

These criticisms chime with Hajer's (2003b) broader statement about the "institutional void" in which contemporary societies are governed. With moves from government to governance, the rules and norms of policy-making are not clear (175). Even if traditional institutional processes are not entirely obsolete, policy-making increasingly occurs within "new political spaces" (178). These are the unstable arrangements by which a range of actors (including consultants for example) make policy, and in the process, the rules and norms of policy-making. The polity is thus not merely the source of policy-making but also an outcome of policy-making – it is increasingly realised in an ad hoc fashion through processes of governance (Hajer 2003b; Hansen and Sorensen 2005).

From such a perspective, the institutional rules and norms of the Berlin political system were determined through the process of privatising BWB. Indeed, the formulation of these rules is as much a part of the neo-liberal reform as the policy itself. This is not to say that traditional institutions or existing modes of decision-making become obsolete. Instead, it is to state that their importance cannot be assumed in the analysis. Consequently, the traditional sources of democratic legitimacy in policy-making can also not be assumed. New forms of governance, as the criticisms of consultants in politics suggest, can destabilise and undermine the traditional rules and institutional practices of representative democracies.

Conclusion: studying facts and values in policy-making

This Chapter has discussed the problems of rationalism in the practice and the study of politics and argued for a move to a constructivist perspective. More broadly, it has also questioned the conventional images of political decision-making. If, then, as a satisfactory model of governance, "orthodoxy is dead" (Miller and Fox, 2007, 3), how should contemporary policy-making be critically analysed? How should the institutions of political systems as well as the claims they make be treated within this context of emerging 'shadow governments'?

The "assumptions and strategies" of many researchers of politics are "more or less rational" (Dryzek 1994, 3). Thus much of the mainstream positivist policy studies literature, like many of the systems they are analysing, rest on a problematic claim to rationality. This has most commonly manifested itself in the positivist, empiricist methods of policy analysis, with its orientation towards quantitative analysis, its objectivist distinction between facts and values and the search for generalisable findings, which hold true regardless of contextual factors (Fischer 2003, 4). As with the practice of politics, objectivism aimed towards the achievement of rational decision-making, conceals in-built biases and assumptions. It conceals the interpretative nature of the work of both policy analysis and policy-making.

Following from this, it has been argued that there is something of a symbiotic relationship between the work of policy analysts and the conventional means of policy-making. Both can be seen to inform a technocratic form of governance in which efficient, informed administration is the best means of making policy. In both, the management of political problems through the deliberations of experts often takes precedence over democratic concerns. In positivist policy analysis there is a "subtle antipathy" (Fischer 2003, 5) toward democratic processes, a dislike of disruptions and the disorder which may arise as the result of disagreements and opposition. This tendency is also apparent amongst politicians and administrators who aim to ensure the smooth implementation of policy (Fischer 2003, 5). As a result, Hajer (2003b) has stated that the mainstream policy analysis literature has helped to reinforce the rationality of politics, frequently becoming a tool of policy-makers. Going further, Fischer has asserted that rational, value-free policy analysis is not only impossible but also serves as an "ideology that masks elite political and bureaucratic interests" (Fischer 2003, 14).

The starting point for a different way of comprehending and analysing politics is to begin by problematising both the notion that there is such a thing as an entirely rational, objective policy-making and that policy will always be made through the formal institutional processes of liberal democracies. It is necessary to adopt a constructivist approach, one that emphasises the contingency of facts and institutional processes. As will be revealed in Chapter 6, such an approach is entirely necessary to get beyond the official accounts of the privatisation provided by key participants. Put in very simple terms, policy moves from being a "self-defining phenomenon" (Fischer 2009, 172) – what will later be termed 'ready-made' – to the contingent outcome of actors constructing relations and producing knowledge.

The post-positivist literature aims to "demonstrate that politics and policy are grounded in subjective factors" (Fischer 2003, 14). The task of the policy researcher is to reveal that what is often presented as the objective truth is more often than not, the "product of deeper, less visible, political suppositions" (Fischer 2003, 14). It is not then possible to draw a clear distinction between facts and values. Claims about the reality of things, that the private sector is more efficient, however, seemingly 'normal' in certain contexts are built upon certain assumptions about the way that reality should be. Policy proposals, however apparently rational or commonsensical, involve making such claims about reality. This is why an appreciation of discourse is important. Discourse analysis "does not look for truth but rather at *who claims to have truth*" and as a consequence we should look at the – explicit and implicit – justifications for claims in policy-making (Carver 2002). There is then a need to appreciate that the production of knowledge is not a neutral process, that there is not one reality 'out there' waiting to be governed (Gottweiss 1999, 63). Rather realities have to be enabled through policy-making.

In politics this does not mean that there cannot be a 'fact' that everyone agrees upon – that Berlin was poor, for example – rather that this may not always be the case. Such facts have to be processed by individuals and it is here that values and theories about the world shape the means through which they are understood and described. The task for the study is, then, to account for the discursive and institutional production of the partial privatisation of BWB.

4. Governmentality, policy discourse and translation

Introduction

Policy-making in Berlin in the 1990s is as much a story of urban governance in the context of globalisation as it is an account of reunification. During this decade Berlin was suddenly exposed to the processes other industrialised cities had been facing since the 1970s. It is a decade defined by policy-making to "re-invent Berlin as a post-industrial service metropolis" (Haeussermann and Colomb 2003, 201), as a 'global city'. As stated in the introduction, this image of the new city was based on the perceived needs of the globalising world economy in the period and the apparent appropriateness of neo-liberal policies to deal with them. These discourses were dominant in the period. But how should their effects on policy-making in 1990s Berlin be conceived?

The example of New York given in Chapter 1 warns against assuming that broader structural and discursive dynamics simply determine policy actions; that urban governance is simply determined by the broader discourse of neo-liberalism. Water governance is intricately connected to local political, environmental and economic processes, as research on cities has revealed (for example: Gandy 1997; 2004; 2005; Kaika 2004; Heynen *et al.* 2005; Swyngedouw *et al.* 2006). Neo-liberalisation is not an "actorless forcefield of extralocal pressures and disciplines" (Peck and Tickell 2002, 53). Local factors are as much a part of the analysis as global factors. Policies such as privatisation are ultimately made by actors. Although the privatisation of the water company was reflective of broader trends of the period, simply referring to these trends tells us little about the process itself or how and why the privatisation took the form it did.

If the post-positivist, interpretivist literature provides the point of departure for the analysis of the BWB privatisation, there is still a need to develop a set of theoretical and conceptual tools with which to analyse the policy process. As stated in Chapter 1, any explanation of policy change rests upon an assessment of the inter-related issues of the nature of the state, political power and the process of policy-making itself (Hill 1997, 41; Wilson 2000, 247). The last chapter placed the rise of privatisation in the context of globalisation and neo-liberalism. This is the intellectual, normative and material context in which the role of the state has been contested and re-configured since the late 1970s.

Despite a general consensus on what the manifestations of neo-liberalism are (see Chapter 2), there is some disagreement on how the effects of neo-liberalism should be understood. How, for example, should the interrelations between political agency and broader political discourses be conceptualised. In particular, how did neo-liberalism come to frame the perceived challenges of globalisation, of re-integrating the city in the global economy? And how did it come to inform water policy-making and ultimately the partial privatisation of BWB? This chapter addresses these issues with further reference to the interpretivist policy analysis literature, particularly Hajer's (1997; 2002; 2003a; 2003b) notion of policy discourses, and two other literatures: Neo-Foucauldian governmentality studies and, from the field of science and technology studies, Actor-network Theory (ANT). It is argued that the governmentality and ANT perspectives provide complementary foci and sets of analytical tools. Governmentality provides a general approach within which to understand government and the role of the state, exposing the constitutive effects of discourses in specific contexts of politics. The ANT notion of translation, on the other hand, focuses the analysis on actions and interactions between actors, seeing them as being generative of meanings. In their different ways, both perspectives are concerned with the ways in which facts are often politically constructed. Furthermore, both share a Foucauldian understanding of power as being an outcome of relations. Power is not possessed and wielded by actors but bound-up in making "authoritative representations" of the world (Jasanoff 2006, 26).

Governmentality

A key concern of Foucault's work was the exposure of the rationalities that informed societal practices – to reveal the means through which societal practices were normalised. If Weber stressed that there was no single rationality, no universal means by which to judge thought, but only a very specific rationalism to emerge in Western industrialised nations, then Foucault sought to reveal that there was a range of rationalities at work in the governance of these nations (Dean 2008, 11). These rationalities apprehend the world in different ways; make it governable in different fashions. Foucault aimed to reveal how seemingly universal and eternal practices were, in fact, constructions specific to very particular times and places. For example, in *Discipline and Punish* (1991), his study of prison systems in 18th century England and France revealed the contingent emergence and normalisation of notions of discipline and control.

Just as Foucault had sought to unearth the modes of thinking embodied in practices of other social realms, such as the prison or the hospital, the notion of 'governmentality' points to the mentalities of governing embodied in political practices. With his simple bringing together of two words, 'government' and 'mentality', outlined in lectures given in 1978, Foucault brought his ideas about power and the constitutive role of discourses to bear on the study of government and the state (Gordon 1991, 1). Governmentality hinges on Foucault's view that the structures and practices of governing can be seen to embody a mentality of governing. In other words, an understanding of the material forms governing takes is impossible without an appreciation of the political rationalities – the modes of thinking and means of reasoning – which have informed their construction. Foucault was referring not only to the modes of thinking inherent to political practices, but to a particular form of government and the political rationalities which shaped its practice. The notion of governmentality represents Foucault's thoughts on the specific configuration of government, which emerged in early modern Europe: the emergence of liberal government, the subsequent separation of the state from society and a new concern with the 'conduct of conduct' (Gordon 1991, 2).

The governmentality literature

Foucault's work on governmentality – and, more broadly on government and the state – did, however, remain undeveloped (Dean 2008, 5; Barry *et al.* 1996, 7). Despite this, governmentality has developed into a key perspective in political science, political sociology and geography, particularly since the early 1990s. Writers in this field have drawn on Foucault's work to come to terms with the rise of neo-liberalism and new forms of governance (Dean 2008, 2). It is to these writers, rather than Foucault, that the study turns in the discussion of governmentality. The aim is not to significantly engage with Foucault's theories or the considerable literature surrounding them but, in more practical terms, to utilise this notion of governmentality to sketch out a general way of thinking about government. This study draws on key texts from the fields of policy analysis (Gottweiss 1999; 2003), geography (Larner 2000; Larner and Walters 2004) and, most particularly, the political sociology of Miller and Rose (2008; see also Rose and Miller 1992) and others (Gordon 1991; Dean 2008; Lemke 2001). More specifically, the aim is to conceptualise the interplay between macro (political rationalities) and micro – local – political practices.

There are actually two meanings to governmentality in the literature (Dean 2008, 16). First, following Foucault's lectures that governmental practices both enable and embody a particular mentality and second that, at least in Western states, shifts in government should be seen as the contestation of a particular liberal form of government. The emergence of liberalism in the second half of the 18th century was seen by Foucault as the emergence of modern government, the designating of limits to governmental power in the shape of a civil society and the market: the constitution of the realms of "freedoms and activities outside the legitimate sphere of politics" (Miller and Rose 2008, 59). For Foucault all politics since this shift (in Western states) could be seen as occurring within the changing debates around liberal government (Gordon 1991, 16). Liberalism did then establish an ethos of government (see Barry *et al.* 1996), a particular way of thinking about the relations between state and society. Foucault argued that discussions of state-society relations in Western states have since centred upon the problematisation of this liberal form of government (Foucault 1991, 102-4).

A more general emphasis on the linkages between thought and practice in government has emerged in the governmentality literature. Governmentality points to the mentalities of governing, which are enabled through practices of government (Dean 2008, 16). Much of the literature, following Foucault, has focused on the constitution of the individual, the links between government and the behaviour of individuals; on the 'bio-politics' of contemporary societies, Foucault's term for the administrative apparatuses that have arisen to govern the health and welfare of society. For example, Miller and Rose (2008) have focused on the co-construction of individual identities (e.g. as consumers) and neo-liberal governmental technologies. It stresses that the thinking involved in political practices is collective and can become seen by individuals as almost commonsensical (Dean 2008, 16). In other words, the rationality informing political practices becomes embedded in the practices of government to the extent that it is not always discernible to those involved in the practices. Governmentalities render a reality, make claims and establish 'truths' about the world. An effect of this is that in particular periods certain policy options resonate more than others. As has been mentioned, in the 1990s neo-liberalism came to have a self-evident, commonsensical quality, being described by Beck as a new ideological "thought virus" (in Peck and Tickell 2002, 34).

It is argued that the governmentality literature provides a valuable approach through which some of the interdependencies between macro (political discourses) and micro processes (political agency) can be explored in policy-making. Foucault's development of the notion of governmentality has been interpreted as being, at least in part, a response to criticisms that his "micro-physics of power" analyses of practices in prisons, schools, hospitals, etc., failed to address macro issues of power relations between State and society (Gordon 1991, 4). However, in addressing government Foucault did not change method, refusing to see the State as having inherent qualities or its own peculiar logic aimed at increasing its own powers at the expense of society (Gordon 1991, 4). Foucault argued that the State is merely a product of the practices of government and not – as usually argued – the source of the practices of government. Thus a focus solely on the institutions of the State is misplaced. Rather, governmental practices – and the mentalities that inform them – define the institutions of the State. The section below draws primarily on a key piece by Miller and Rose (2008; Rose and Miller 1992), *Governing beyond the state: problematics of government.* In this the authors, drawing on Foucault's work on governmentality, propose three dimensions to the study of governing beyond the state: 'political rationalities', 'political programmes' and 'technologies of government'.

Political rationalities, political programmes and technologies of government

Political rationalities

Governing, the projection of political power, can be analysed in terms of the fundamental contests between "political rationalities". These are the broader discursive frames with which different forms of political power are morally and theoretically justified: the means by which those who govern are identified and the objects and limits of politics are defined (Rose and Miller 1992, 175). Political rationalities, such as Keynesian and neo-liberalism, are the broader notions of what the State is and what it should do. They have a moral vein running through them: they seek to set the boundaries for the correct exercise of power, the limits of authority, the duties of government and the proper distribution of tasks amongst societal and state institutions (Miller and Rose 2008, 58). In their attempt to fix the moral objectives of government – to ensure prosperity, safety, etc. – political rationalities have an epistemological character. They evolve in relation to particular conceptions of the objects to be governed – the economy, the people and so on (Miller and Rose 2008, 58). Finally, political rationalities are expressed within particular idioms or vocabularies which, being more than mere tools of rhetoric, serve as a kind of "intellectual machinery" to make the world comprehensible (Miller and Rose 2008, 58). Political rationalities are then "morally coloured, grounded upon knowledge, and made thinkable through language" (Miller and Rose 2008, 59).

Political programmes

It is important to note that such rationalities do not simply create political realities. Rather political rationalities inform the ongoing process of governing. "Government is a problematizing activity: it poses the obligations of rulers in terms of the problems they seek to address" (Miller and Rose 2008, 61). In response to these problems political programmes are developed. With reference to management of the economy, typical neo-liberal programmes include privatisation, liberalisation and commercialisation.

Political programmes problematise aspects of state-society relations; be that the state and the economy, the state and the environment or the state and education. Importantly, as Rose and Miller point out, political programmes always rest upon claims to knowledge about the field in which such programmes are to be rolled out. They make a claim about the way the world is, and in the process of making this claim, also make it pliable to practices of government. Proposing the privatisation of a water company involves making claims about the performance of a public company. It is about questioning the 'realities' of public ownership and proposing the means to improve upon it. Thus political programmes define the scope and duties of government, objectifying aspects of the social and designing the means to transform them (Rose and Miller 1992, 181).

> "Governing a sphere requires that it can be represented, depicted in a way which both grasps its truth and re-represents it in a form in which it can enter the sphere of conscious political calculation" (Rose and Miller 1992, 182).

From this perspective, privatisation programmes are about the production of new knowledge. In the water sector, privatisation rests upon the creation of new facts about the appropriate role of government in the provision of WSS services and the most efficient way of managing them: the way in which the water sector should be seen to have operated under public control and, crucially, the ways in which it will operate after privatisation.

Governing is then a problematising activity. It is constitutive of a reality: it outlines an object of government and proposes a means by which it should be governed. The constitution of this reality, say the performance of a public education system, the government's finances, its ambitions and roles in the education sector, are not based on neutral knowledge, nor are the solutions proposed to reform this reality. They are instead rooted in moral, technical and theoretical assumptions about the world. Programmes such as privatisation, are informed by political rationalities (in this case neo-liberalism) but they are not determined or derived from them. Instead, Miller and Rose (2008) refer to the term 'translation', developed by Actor Network Theorists, to conceive of the relationship between programmes and rationalities. To Miller and Rose (2008) translation entails "both a movement from one space to another and an expression of a particular concern in another modality" (60). Political programmes are therefore contingent upon their translation in a particular place at a particular time.

From a governmentality perspective, the BWB partial privatisation is not merely the product of neo-liberal ideology, or a pragmatic response to the economic situation of Berlin in the 1990s. Instead, the privatisation should be seen as the contingent outcome of global-local dynamics: the context-specific translation of neo-liberalism. It is, furthermore, part of a more subtle "political project that endeavours to create a social reality that it suggests already exists" (Lemke 2001, 13). To emphasise this key point, a political rationality is not neutral knowledge that "simply re-presents" reality, rather it is constitutive of the reality which governmental practices are linked to (Lemke 2001, 2).

Experts

Expertise is fundamental to these processes of representation and the linking of government practices to them. Experts (scientists, economists, bureaucrats, private sector consultants, etc.) provide the "intellectual machinery" of governing, creating the means (through their theories, concepts and models) to make the world first thinkable and then manageable (Rose and Miller 1992, 182). Importantly, experts work with a "language claiming the power of truth" (Miller and Rose 2008, 68), removing problems of government away from the "disputed terrain of politics" to the "tranquil yet seductive territory of truth" (Miller and Rose 2008, 69). It is their expertise and "know-how" (Dean 2008, 23) which can provide politicians with an aura of authority, and reinforce government claims to objectivity and truth. This is why experts are at the heart of de-politicisation processes. They can, however, contest political authority. Potentially, experts can become so integral to government that such de-politicised arrangements are enclosed from other actors in government (Rose and Miller 1992, 188), as the literature on consultants in politics suggests. In the BWB case it is clear that consultants are bound-up in the processes of programming and operationalising neo-liberalism: given their knowledge of the private sector, they are essential to politicians who wish to privatise, liberalise, commercialise, etc.; given their interests, they are also, of course, likely to further encourage the rolling out of neo-liberalism.

Technologies of government

Miller and Rose's final dimension to an analysis of governing is that of 'technologies of government'. This refers to the means through which political programmes are rolled out in a particular field; the "strategies, techniques, and procedures" through which the actors attempt to make their programmes operable (Rose and Miller 1992, 183).[15] According to Miller and Rose (2008), the list of technologies of

[15] In the field of political science, see Barry 2001 and Walters 2004 for discussions of the technologies of EU integration.

government is "heterogeneous and in principle unlimited" (63). Examples they give include: "techniques of notation, computation and calculation; procedures of examination and assessment; the invention of devices such as surveys and presentational forms such as tables" (Miller and Rose 2008, 63).

Although a concern for the technical aspects of government is apparent in the governmentality literature, government is not seen as being reducible to these aspects (Dean 2008, 31). Rather, technologies of government should be understood in a broader sense as the "complex assemblage of diverse forces" that enables the deployment of political rationalities and the political programmes which articulate them (Miller and Rose 2008, 63). Adding clarity to this dimension, Dean (2008, 212) refers both to technologies and techniques of government in his discussion of the range of means through which governing is accomplished.

Technologies of government are therefore the context-specific means through which broader political rationalities and programmes are translated into local practices of government. Referring again to ANT, Miller and Rose argue that this process should be seen as being one akin to translation rather than implementation. By this they mean that an ideal type of privatisation is not merely implemented by force from the centre of government. Instead, privatisation or any other programme should be seen as the bringing together of legal, administrative, political and societal actors into a network. Following the ANT view of power, itself in part Foucauldian, political power is seen as being an effect of such assemblages, rather than the cause of them. Miller and Rose argue that the power of a government is not simply the result of their institutional privileges. Rather it is the ability, at a particular time, to "successfully enrol and mobilise persons, procedures and artefacts in the pursuit of goals" (Miller and Rose 2008, 64). Realising political programmes is about forging shared interests out of diversity (Miller and Rose 2008, 64).

From such a perspective, policy-making is a process full of contingency and compromises, in which actors attempt to enrol others in these governmental networks. Political rationalities may inform practices of politics, but the emphasis in this approach is on the interplay between the three. They need to be realised through programmes of government and mobilised through technologies of government (Rose and Miller 1992, 181). Neo-liberalism should not be seen simply as an abstract political philosophy but rather an outcome of the interactions between broader political rationalities and the local institutions, organisations and actors assembled in governmental technologies to pursue particular objectives.

Seeing neo-liberalism as a governmentality also moves us beyond notions of it as an ideology. It requires the researcher to see discourse as being not only the outcome of power relations but actually constitutive of them. Discourse is understood not merely as:

> "rhetoric disseminated by hegemonic economic and political groups, nor as the framework within which people represent their lived experience, but rather as a system of meaning that constitutes institutions, practices and identities in contradictory and disjunctive ways" (Larner 2000, 12).

Such an approach places the production of knowledge at the heart of governing. Governing centres on processes of knowledge production, through which entities are constituted as 'political', and then linked to programmes and technologies of government. Governing may be underpinned by rationalities such as neo-liberalism, which provide notions of what is good, normal, efficient or profitable but it is a "domain of cognition, calculation, experimentation and evaluation" (Miller and Rose 2008, 55).

Consultants and governance

What are the implications of adopting a governmentality perspective for the analysis of consultants and how does this differ from the existing literature on consultants and government? Much of the debate in the literature regarding consultants hinges on whether they are forcing themselves into politics or whether politicians are actively recruiting them; whether the rise should be explained with reference to broader structural trends or their individual actions and motivations. Related to this, is the question of how much influence consultants actually have on policy processes: whether they are merely providing specialist, technical advice or whether they are in fact making political decisions. Are consultants furthering their own interests or are they bolstering the power of the politicians who hire them? Are they someone to take the blame for politicians if things go wrong (as a consultant working for the RWE/Vivendi/Allianz bid stated – Interview: 8)?

One problem with such discussions is the attempt to clearly distinguish between the technical and the political in policy-making. Grindle's (1977) study of the rise of the *technico* (the rational, technical expert) in the Mexican bureaucracy in 1970s contradicted the accepted view in Latin America that politicians of influence monopolised political skills while the bureaucrats of influence monopolised technical information. In reality, to have a real influence on decision-making, bureaucrats needed both political and knowledge-based skills (402). Bureaucrats of influence were not mere providers of technical knowledge. Knowledge was not enough; it had to be politically deployed. Saint-Martin (2000) makes a similar point: consultants, along with the reforms of government they help realise, have political as well as technical objectives (Saint-Martin 2000, 198).

If it is not easy to distinguish between the technical and political, a broader problem with the consultant focused literature is that it risks reifying the importance of consultants and, in the process – ignoring the complexity of policy-making. Though a focus on consultants has the advantage of rooting the analysis in the concrete actions of consultants and governments, it suffers from a tendency to over-estimate the rationality of consultants; the extent to which they can simply realise their – financial and political – interests in the competition to influence governmental practices (Saint-Martin 2000, 194).

Furthermore, ascribing an 'interest' to a type of actor simplifies the agency of consultants in politics. In the case of consultants in the BWB partial privatisation, they were not merely working for their own interests, or the private sector interests, against the traditional public sector actors. There were consultants working for the interests of the Berlin government. It is therefore difficult to conceive of them as a single type of actor. It is furthermore, difficult to research them, particularly given the sensitivity of their roles in the BWB privatisation. In the Berlin case research was unable to trace a causal relationship between the lobbying of consultants and private companies and the political decision to privatise.

Understandably, the debate on consultants is highly charged and should remain so, given the issues at stake for democracies. Thus research should aim to uncover cases where consultants do manipulate the policy process for their own – and others' – interests. However, approaching the subject in an instrumentalist fashion does not bring the researcher any closer to conceptualising the inter-dependencies between reforms of governmental practice – and decision-making – and the increasing importance of these experts. This study argues that a governmentality perspective provides the foundations for a more productive way of thinking about consultants and neo-liberalism. Thinking of experts as being fundamental to particular modes of governing, as actually enabling forms of governance, merges any assessment of consultants with an assessment of neo-liberalism itself. The crucial move here is to 'de-centre' the consultant – to analyse them within the discursive context within which they act. Utilising Miller and Rose's approach, private sector consultants should then be conceptualised as operating in-between the broader discourse of neo-liberalism, the programmes designed to achieve it and the material political practices which emerged in 1990s Berlin.

The key to understanding government as governmentality is not to study neo-liberalism in the abstract, as an ideology or political philosophy, but to concentrate on the means through which it can become embedded within practices of government. To "analyse mentalities of government is to analyse thought made practical and technical" (Dean 2008, 18). The concern in this study of the BWB privatisation is therefore to reveal how a water company became re-fashioned within a mentality of government: how it became problematised – re-represented within this mentality of government. The aim will be to show how BWB was re-made in the terms of neo-liberalism – how knowledge was produced about it: how it became subjected to and tangled-up within, the practical, technical means of establishing this mentality of government.

Consultants were integral to the re-making of BWB. It is shown that their expertise in this privatisation, as well as providing them with integral roles in the production of facts in governing, also endowed them with crucial managerial roles. Indeed, such is their importance to the privatisation process, and the elite politicians leading it, consultants can be seen to create what Rose and Miller term 'enclosures' within it: "relatively bounded locales or types of judgement within which their power and authority is concentrated, intensified and defended" (Rose and Miller 1992, 188). This authority is established through the deployment of "esoteric knowledge, technical skill, or established position as crucial resource" (Rose and Miller 1992, 188): their knowledge of the private sector, of how privatisation deals 'should' be done, in other words, the very qualities on which politicians depend in privatisation deals.

Consultants in this sense can thus reinforce political authority. A politician's, or party's authority 'externally', in relation to other political parties may be bolstered by high profile experts, actors within government, the broader public, etc., while their actual dependence on these experts for policy proposals, media tactics, and so on reveals this authority to be a shared resource. Indeed, politicians can be excluded from crucial processes of knowledge production – despite colluding in their formation – because they do not have the requisite knowledge resources.

Overall, it will be suggested that consultants are emblematic of the neo-liberal re-definition of the state's role in water management in Berlin. More generally, it will be suggested that, if the Keynesian state represented the growth of state bureaucracy and the pre-eminence of the bureaucrat as a manager of government, then the neo-liberalised state can, perhaps, be defined in terms of the spread of the market and the increasing pre-eminence of the consultant as the manager of governance. In the case of Berlin, the replacement of one language 'claiming the power of truth' (that of the state as arbiter) with that of the market (as a model for governing) leads, in turn, to a shift in the language and logic of governing.

Summarising government as governmentality

The governmentality literature draws together Foucault's thinking on how power is 'effected' and the generative effects of historically and geographically specific discourses. From a governmentality perspective, governing, the projection of political power, becomes less an account of rational actors or rational actions and more an account of how political rationalities inform practices of politics and how, in the process, they become transformed through these practices. In line with Foucault's more general thoughts on how social structures emerge, governmentalities are seen as emerging through actual practices of governing. According to Miller and Rose's analytics of government, political rationalities may inform actions – provide a moral and theoretical compass – but actual mentalities of governing are enabled only through action – the compromises and contingencies of politics practiced in specific contexts.

The aim of this approach – and more generally the governmentality literature – is to trace the developments of such technologies, revealing their emergence from the contests between broader political rationalities such as neo-liberalism and Keynesianism and reveal the contingent construction of specific webs of relations and rules through which society becomes governed: the contingent means by which neo-liberalism is brought to bear on realms within society. Through such work the aim is to reveal the contingency of governance; by showing how the reliance of even grand governance projects, on very specific links between actors and governmental technologies the aim is to reveal their fragility, to present them for re-contestation. The exercising of political power does not then become inevitable, the implementation of the wishes of an elite driven by ideology; rather power is seen as being contingent on the mobilisation of a range of institutions, organisations and individuals. Neo-liberalism, being realised only through the construction of governmental technologies and the translation of governmental networks becomes less coherent, more contingent and therefore open to contestation.

> "This politics stresses the complexity, ambiguity and contingency of contemporary political formations to maximise possibilities for critical responses and interventions" (Larner 2000, 14).

Adapting the approach to analyse policy-making

The intention of Rose and Miller as well as the literature more generally is to reveal the interdependencies between thought and practice in a particular place at a particular time: to emphasise the contingency of government. As useful as this approach is, Miller and Rose's approach remains an analytics of government (2008,

54) and not a framework for the analysis of policy-making. Much of the governmentality literature, following Foucault, has focused on the links between government and the behaviour of individuals; on the 'bio-politics' of contemporary societies, Foucault's term for the administrative apparatuses which have arisen to govern the health and welfare of society. For example, Miller and Rose (2008) have researched the co-construction of individual identities ('consumers', for example) and governmental technologies.

Larner (2000, 14), however, has criticised the literature for failing to undertake the kind of research that it claims is necessary: delving into the "messy actualities" of specific neo-liberal programmes. Instead, through focusing on broader governmental themes such as welfare or health reform (see chapters in Miller and Rose 2008), governmentality researchers run the risk of doing what they specifically aim to avoid: "producing generalised accounts of historical epochs" (Larner 2000, 14).

Such criticism is not entirely fair as a later assessment by Larner and Walters (2004, 3-4) seems to indicate. Miller and Rose (2008) have, for example, provided fairly detailed accounts of one of their chosen empirical objects, the Tavistock Clinic in the UK and Barry (2001) illustrated his account of the EU's governance of "technological zones" with a number of carefully chosen case studies. Nonetheless, this work is certainly not in-depth, 'thick' in its empirical description (McKee 2009, 473-4).

To avoid such a criticism, while in the process adapting the approach for the analysis of policy, this study takes two steps. First, the approach needs to better account for, or at least pay more attention to, the interplay between discourse and agency. To achieve this, a more explicit focus on political agency and the institutional context of politics is required. It is argued that a more thorough utilisation of the notion of translation provides the means to do this. The second step involves addressing the issue of broader discourses and context-specific discourse. It is argued that the mid-range conceptual tool of policy discourse, as developed in particular by Hajer, provides a means of understanding how political rationalities are distilled into specific policy discourses.

In the case of Berlin, it offers a better means of thinking about how the global came to mix with the local in policy-making in 1990s Berlin: how global discourses such as neo-liberalism localise. Issues of macro-micro and structure-agency are, of course, at the centre of social sciences debates. By addressing them, the study reveals the tensions in the approach itself. These are not ultimately resolved. Rather, the study aims to utilise the different approaches to better explore the privatisation policy process.

As stated, in the 1990s, many city governments were privatising WSS services in response to the perceived challenges presented by globalisation. However, any account of policy change in WSS which referred only to these global trends as a means of explaining what happened in Berlin would be incomplete. The actions of policy-makers and other actors in Berlin cannot be divorced from the local political context. The specific manner in which these global trends were apprehended in Berlin, and the way in which governmental practices were re-configured, were shaped by local factors.

In the 1990s Berlin re-gained its capital status in the reunified Germany and the notion emerged that Berlin should and would become a 'global city'. In the case of Berlin, this sense of what the city represented, or what it should become or was already becoming, was crucial to water policy-making and urban governance in general. As stated, Berlin became a symbol for reunification and the wider changes of the 1990s (Cochrane and Passmore 2001, 343). The approach of Miller and Rose provides a general means of conceptualising the interplay between macro and micro, discourse and agency. But where would a discussion of the importance of the history and symbolism of Berlin fit into the approach? Miller and Rose provide the notion of political programmes as the means through which rationalities are translated into particular contexts of political action. This is useful as a general way of thinking about the relations between neo-liberalism and policy-making in a particular context. What it absent is a set of conceptual tools to pick apart the more specific discursive dimensions of a policy process: a means of analysing the discursive resources employed by actors in a particular context. For this reason the study returns to the interpretivist policy studies literature and in particular Hajer's (2003a) conceptualisation of policy discourses.

Hajer's terms of policy discourses

Hajer's (2003a) discourse analysis perspective centres on the examination of "the terms of policy discourse" by which he means concrete "policy vocabularies", "story lines" and "generative metaphors" which shape the positions actors take (103). From this perspective, beliefs are sublimated in the terms of policy discourses which, taken together, account for the ways in which "biases are structured in textual utterances" (Hajer 2003a, 104). Policy discourses do then have a similar quality to broader political rationalities. They work to rationalise a particular form of political power, to provide theoretical and moral arguments for a configuration of state-society relations. As Gottweiss (1999) states, however, policy discourses or "policy narratives" are more specific than meta-narratives such as neo-liberalism in that they "describe the frames or plots used in the social construction of the fields of action for policy-making" (64).

Within Hajer's conceptualisation of policy discourses, the language of political rationalities may not be to the fore. Rationalities are more likely to be embedded within the terms of policy discourses. In other words, they shape the terms of the discourse, but they may also be concealed within them. A policy discourse may not be overtly neo-liberal in its language and tone even though its assumptions, strategies and ultimate objectives are. In the EU, for instance, policy discourses surrounding the Emissions Trading System (EU ETS) are rarely explicitly neo-liberal. Rather, the assumptions and strategies of neo-liberalism are located within policy language concerning Greenhouse Gases and Climate Change. Of course, the logic and desirability of the market are present, but it is translated into the context of environmental governance.

Similarly, in 1990s Berlin, the global city policy discourse was not overtly neo-liberal. Rather, it is in the framing of the challenges posed by globalisation, the qualities Berlin was deemed to have and the policies it had to implement that neo-liberalism can be discerned. Neo-liberalism did not just inform the policy-making means through which Berlin could become a prosperous city. It shaped the vision of what kind of city Berlin should become, within the context of economic globalisation.

Thus Hajer provides a set of analytical concerns to explore the discursive means through which rationalities can become normalised in specific policy-making contexts – the ways in which neo-liberalism can become embedded through storylines, policy vocabularies and generative metaphors. Additionally, a concern for the terms of policy discourses has the advantage of focusing the analysis on the discursive work conducted by actors in a specific place.

Within these terms of policy discourses, it is the notion of the storyline, which is most fundamental. Drawing on metaphors and myths, storylines serve to "sustain societal support for particular policy programmes" (Hajer 2003a, 104). They hold policy coalitions together through functioning to "'interpret' events and courses of action in concrete social contexts" (Fischer 2003, 102). In other words, we can see such storylines as providing something akin to a shorthand for beliefs, as they "symbolically condense the facts and values basic to a belief system" (Fischer 2003, 102).

Storylines become particularly powerful, and act as a catalyst for change, when they succinctly "bring together previously unrelated elements of reality" (Hajer 2003a, 104). At the start of the 1990s, it was unclear how the reunified Berlin, the reunified Germany, the globalised economy, the EU and the rapidly changing Central and Eastern Europe related to one another. A storyline that 'Berlin is becoming a global city/ Berlin must restructure to become a global city' stressed that Berlin had to restructure its economy because of the pressures of globalisation and that this restructuring would inevitably lead to Berlin becoming a service-based economy. There were then genuine structural pressures on Berlin to become a global city.

Neo-liberal globalisation is not understood solely in discursive terms: it had clear material effects. Rather it is argued that to fully understand how it shaped policy-making, it is necessary to trace the ways in which it was discursively produced in Berlin. As Gottweiss (1999, 63) states:

> "the 'truth' of an event, a situation or an artefact will always be the uncertain outcome of a struggle between competing language games or discourses which transform 'what is out there' into a socially and politically relevant signified".

Global city discourse

The key to understanding the effects of economic globalisation in Berlin is then about much more than assessing economic performance or structural pressures. It is about revealing the ways in which they become embedded and produced through the terms of discourse. This storyline translates the broader context of globalisation and neo-liberalism into a Berlin-specific policy discourse: it provides the 'New Berlin' with a sense of purpose and a place in the world. This storyline provided politicians, businesses, Berliners, the outside world in general, with a means with which to "'interpret' events and courses of action in concrete social contexts" (Fischer 2003, 102).

Berliners and many of their policy-makers may not have understood the full – financial, economic, institutional and social – implications of policies aimed at creating this global city. However, this storyline offered a simpler means of comprehending them: a frame through which to make sense of the changes occurring during the decade. In other words, storylines allow people something upon which to focus, ruminate and make strategic decisions. In this approach it is assumed that most actors, outside of professional experts (scientists, policy specialists, etc.), who are more focused on the assimilation of evidence, respond to "simplified storylines" reflecting the "concerns of core beliefs, rather than the beliefs themselves (Fischer 2003, 102). Through such simplification, storylines are potent because they "help people to fit their bit of knowledge, experience or expertise into the larger jigsaw of a policy debate" (Hajer 2003a, 104).

To justify the importance of storylines, Fischer draws on the social psychology literature to argue that most people's beliefs are not just the outcome of intellectual reflection – emotions and feelings also play a role. For instance, many people refer to their social groups when defining themselves and have a desire to belong to one group and maintain distance from others (Fischer 2003, 103). Beliefs are thus seen to be the result less of a cool, detached assessment of the facts, than the result of a complicated process, a reflection of "complex social allegiances and animosities" (Fischer 2003, 103). This is not a positivist conception of the autonomous actor, making more or less rational decisions: individual beliefs and preferences are more fluid, open to emotional appeals. It is within this conception of individual preferences, that storylines can 'stick', providing, as they do "social orientation, reassurance, or guidance" (Fischer 2003, 103). It is probable that the global city policy discourse garnered much of its potency from its attractiveness – its emotional appeal – as a vision for the New Berlin.

The allure of the global city image is understandable in Berlin. It provided a way of escaping its 20th century history of ideological extremism of fascism and communism – to get away from its own past and embrace a non-ideological, even apolitical future. Re-fashioning Berlin as a prosperous, 'high tech' economic and cultural city, open to the global flow of capital, goods and people, would move the city away from the Cold War, the Second World War and the Nazi period and closer to the more flattering image of a 1920s Berlin.

> "The once heavily bordered city of the cold war era now wished to embrace the opposite image of unfettered, borderless, global city-ness, hankering after an electronic-age equivalent of the world city label that Berlin of the 1920s not only attained but exemplified" (Ward 2004, 240).

Beyond this symbolism, the global city policy discourse promised prosperity. It captured Berlin's aim not merely to become 'normal' but the belief that Berlin had the potential to be successful in the globalising system of free markets and democracies. Optimism about Berlin's future rested in part on its location and the opening-up of markets in Central and Eastern Europe. The question of Berlin's capacity to restructure and re-develop to become a city of global importance became increasingly connected to the perceived advantages of its location within the EU as enlargement into Central and Eastern Europe progressed (Kraetke 2001, 1777). Indeed, by the end of the decade, the rhetoric had shifted a little, from Berlin becoming a 'European city' instead of a 'global city': of becoming a key city within the "European Community of *metropoles*" (quoted in Cochrane and Passmore 2001, 345).

Actors gravitate around such storylines even though they are, given their simplifications, often vague on the 'facts' and even contradictory at times (Fischer 2003, 103). In Berlin, there was an apparent contradiction between the 'becoming a global city' storyline and increasing economic and social problems. However, most actors do not have a detailed knowledge of policy issues to empirically compare the facts to their belief systems. Hajer (2003a, 104) argues that it is precisely the persuasive flexibility of storylines which gives them their potency. In Berlin, the storyline centred on necessary restructuring and adaptation to the global economy worked to justify such apparent contradictions, framing them as necessary hardships in a process which would eventually bring prosperity.

Potent storylines do then have an ontological flexibility, they 'stick' when they are flexible enough to appeal to people who have different core beliefs.[16] Such flexibility was crucial to the global city policy discourse in 1990s Berlin. In other words, the Berlin of the 1990s may still have borne many of the scars of the Cold War and was not yet a key node in the global market economy, but it was becoming one. It explained and justified the problems Berlin was enduring whilst providing the means to solve them. The vital part of the storyline was the sense that achieving the goal was inevitable. In fact, Berlin was already on the way – it was *becoming* a global city.

If storylines are at the core of policy discourses, they are supplemented by other layers of language and meaning which add meat to the bones. The second layer is made of 'concrete policy vocabularies', by which Hajer means the "sets of concepts structuring a particular policy, consciously developed by policy makers" (2003, 105). Hajer's examples from the field of environmental policy-making include 'ecological corridors', 'nature development areas' and concepts borrowed from sciences such as 'natural balance' (2003a, 105-6). The global city discourse consisted of a vocabulary centred on terms such as 'competitiveness', 'high-tech', ' new media', 'the service economy' and so on. Such vocabularies are important as they become chess pieces in debates, strategically placed by policy-makers in their attempt to help set the terms of the debate and win support from other actors.

The third layer to policy discourses is that of "generative metaphors" (Hajer 2003a, 106). By this Hajer means reference to, or inclusion of ideas, concepts or symbols which somehow fit with a time; which have a particular resonance in a particular period of time. An example of this would be 'network', a metaphor for our times, providing us with a way to grapple with the complexity of the world around us, though we usually take its exact meaning for granted when we use it (Hajer 2003a, 107). In 1990s Berlin, the notions of the 'service metropolis' and, even more potently, the 'global city' had such qualities: they fitted, resonated in this particular period of time.

Taken together this policy discourse mobilised bias in policy-making (Hajer 2002, 62), it aimed to discursively produce 'facts' about Berlin's economy and its role in the world: "each policy discourse comes with its won power effects as it shapes the knowing and telling one can do meaningfully"(Hajer 2003a, 107). An important point to remember for the analysis is that "discourses are not static and do not take power by themselves" (Hajer 2003a, 107). Instead they garner potency

[16] Hajer (1997) demonstrates this with reference to the ecological-modernisation discourse's capacity to appeal across perceived divides, uniting actors from the worlds of business, environmental movements and politics.

through 'processes of mutual positioning': "the ways in which actors inter-subjectively create and transform political conflicts using language" (Hajer 2003a, 107). The Berlin case study will then, consider some of the ways in which actors – willingly or unwillingly – were positioned by the terms of the global city policy discourse: by the need to create a competitive, high-tech service-based economy. This process of positioning is not just about "cornering one's opponents in concrete discursive exchanges" but is more powerful in that a discourse can potentially exclude or marginalise certain ways of seeing, of understanding an issue (Hajer 2003a, 107).

The Berlin 'boosters' were, at least at the start of the decade, a large group – Berlin politicians, consultants advising them, local, national and international investors as well as academics and politicians outside Berlin. Many believed it was becoming such a city and much of urban policy-making in the decade was guided by this image of Berlin. Policy-making is ultimately characterised by attempts to make this – neo-liberal – representation of the city a reality.

Though it never quite happened in the 1990s (and still has not), at the same time it never seemed quite to fail. To paraphrase the famous quotation from German writer Karl Scheffler (1910, 267 in Costabile-Heming *et al.* 2004, 3), Berlin was always becoming, never being.[17] This was the key conceit within the storyline. Berlin may not have been a global city but it was on the cusp of becoming one; it was an "up-and-coming" global city (Kraetke 2001, 1777). The boosterism was about creating the 'New Berlin' so that people, particularly external investors and tourists, would believe it (Haeussermann and Colomb 2003, 210).

As the next chapter reveals, the Berlin economy did develop in certain sectors but the overall objectives of restructuring remained far off. Still, the boosters of Berlin ensured the global city image endured through the 1990s as the manufacturing industry collapsed, unemployment grew and the population declined. The gap between image and economic performance (as well as social standards) remained but the path chosen was never obviously contested by the government and their closest advisers. This was tribute to the potency of neo-liberal globalisation, the creation of 'necessities' both discursively and materially: the representation of the world in terms of the market and the simultaneous re-making of the world according to these terms. The difficulties of achieving this global city aspiration were, however, inextricably linked to the Cold War period and the loss of pre-Second World War economic functions to other West German cities.

As will be shown in the next chapter, the global city storyline obscured the extent to which Berlin and the German economy had changed in the Cold war period. The big cities of West Germany had become key centres of industry and

[17] "Berlin is a city condemned forever to becoming and never to being": "*Berlin ist eine Stadt, verdammt dazu, ewig zu werden, niemals zu sein*".

finance during the Cold War period but were glossed over in the global city discourse. The storyline's emphasis on adapting to the perceived needs of the global economy, and employing neo-liberal policies to achieve it, masked the very specific, structural obstacles to becoming a global city.

From discourse to political agency: ANT and translation

Discourse analysis approaches argue that discourses are constitutive of power but there are, of course, different emphases. Hajer (2003a, 103-4) rejects a narrower focus on linguistics alone and favours instead an approach which includes the "institutional practices within which discourse is produced". Hajer (2003a, 120) stresses that analysing the terms of a discourse is not enough, that researchers must look at the coalitions of actors which support policy discourses and the institutional contexts in which they operate. In other words, there comes a point when the analysis must move beyond the assessment of language alone and look at the influence of institutional rules and settings in shaping discourse. Importantly, Hajer notes that "rules and settings are not only immanent in language; it is also relevant to examine the settings in which discourse gets produced" (2003a, 108).

Hajer's discourse approach does then pay attention to the institutions of political systems. It aims to combine an examination of the discursive production of 'truths' with an analysis of the institutional context in which they take place. Writing from a similar perspective, Gottweiss (2003) states that any policy analysis should pay "great attention not only to the organisation of politics, but also to the politics of organisation, not only to the actors of politics, but also to the politics of actors" (254). There is then an appreciation that socio-political practices and strategic interactions are integral to the production of truths and the making of policy. Ultimately, however, the tendency in Hajer's approach, and more generally, neo-Foucauldian discourse analysis approaches, is to explain policy change with reference first to the discourse itself and then to turn to a consideration of interactions between political actors (Rutland and Aylett 2008, 631).

The governmentality literature has also been criticised for a tendency to focus on discourse at the expense of a detailed consideration of how governmental practices are actually defined in a particular place. O'Malley *et al.* (1997) noted that many writers in the governmentality field have studied political rationalities almost in the abstract, apart from the material practices of governing which they inform. One implication of adopting such an approach is that the inter-dependencies between thought and practice – the apparent analytical focus of the literature – becomes somewhat assumed. This is arguably a tendency not only in the governmentality literature but more broadly within Foucauldian discourse analytic approaches: "discourses, so to speak, perform and instantiate themselves. There is nowhere else to go. Nothing else animates them. There is no puppeteer. Instead, they animate themselves" (Law 1994, 107 in Rutland and Aylett 2008, 631). The criticism is that discourses end up explaining themselves – they have their own power – and analyses allow little room for political agency.

Returning to an assessment of Miller and Rose's approach, such an outcome would certainly contradict their emphasis on translation. Translation is crucial to the approach – it is the means through which the inter-relationships between rationalities, programmes and technologies are conceptualised: the relationships between the macro and the micro and discourse and agency are understood. Indeed, one reason Miller and Rose's approach is attractive is that it seems to offer a way of combining an analysis of broader discourses and local level interactions. However, as this chapter goes on to discuss, processes of translation are not a fully developed concern in their analytics of government. The authors stress the contingency of political power, its dependence on aligning a variety of forces. They argue that rationalities and political programmes take on different, localised meanings in these translations. However, the contingency of these processes of translation is not the focus of their analyses. Instead, the priority often lies in tracing the links between rationalities and the thoughts and actions of individuals: the focus is on discursive effects, rather than the means through which these effects are realised.

Ultimately the crux of their approach, that governmentalities are the outcome of the interplay between macro political rationalities, political programmes and local governmental technologies, is never fully developed. The crucial points at which discourse and agency and the macro and micro meet are merely suggested. As such, it is argued that the approach needs to be adapted to offer a more detailed consideration of how such political rationalities relate to political agency. Further elaborating upon the notion of translation allows for a greater concern for political agency and more generally the institutional settings in which policy is made. It provides a better sense of the contingencies of governing. Thus, if the governmentality literature places more emphasis on the 'politics of organisation', then the ANT literature and the notion of translation focuses our gaze on the 'organisation of politics' and ultimately the processes through which political power is effected.

Regardless of whether such criticisms of Hajer and other Foucauldian analyses are fully justified, to really shift the analysis onto the actors of politics, a change of approach is required. ANT is founded on the assumption that social structures, political power and discourses are an achievement of ongoing social relations. From such a perspective, the potency of neo-liberalism had to be achieved in Berlin in the 1990s – it had to be performed, i.e. the implementation of neo-liberal policies is not explained simply with reference to the global dominance of neo-liberalism as a discourse or the structural effects it had. It was an outcome or effect of actions.

Overview of ANT

ANT is an innovative, if provocative approach.[18] Emerging in the field of science and technologies studies, it has proved to be increasingly popular across a range of disciplines including organisation and management studies, geography and political science (Fine 2005). Though it has not made huge inroads into political science, it has influenced the work of a small number of interpretive policy analysts, notably Freeman (2007), Gottweiss (1999; 2003), Gomart and Hajer (2003) and Hajer and Versteeg (2005), who drew on ANT to provide an account of knowledge production as performance.

In general terms, ANT has been utilised to understand the processes through which social structures appear and take root, how they are contested, replaced or sustained over time and distance. This is to be revealed by looking at the construction and expansion of actor-networks. The premise of the actor-network notion is that all action is contingent; that agency is an effect of the sets of relationships, or networks, within which actors are embedded. From an ANT perspective, "one can only act with others; behind every action lies a network" (Murdoch 2003, 269). In ANT, action is not only contingent on arrangements with other people, but also technologies and natural entities, which are seen as being potentially active in the formation of social structures. Callon *et al.* (2002) and Latour (1993) have argued this point by showing how contemporary societies are increasingly characterised by the entanglement of the natural, social and technological, to the extent that hybrid forms of agency increasingly define our society. Actor-networks are, then, heterogeneous assemblages (Law 1992; Latour 2005) of human and non-human entities which are brought together and ordered so as to perform specific actions.

Translation

The notion of translation (adopted by Rose and Miller) is used in the ANT literature to capture the processes through which actor-networks emerge and are stabilised: to reveal how actors, objects and organisations are brought into alignment to achieve particular objectives. Translation, or the sociology of translation (Callon 1986; 1991), emphasises the displacements of people, objects and meanings and the necessary transformations which occur in this process (Callon 1986, 223-224). On

[18] For an overview of criticisms of ANT see McLean and Hassard 2004.

the one hand, then, translation is about the negotiation of meanings, the roles actors play and their relations to each other and the objects around them. Translation is also fundamentally about revealing how agency is realised – how priorities and courses of action emerge from negotiations between actors. As entities cannot act in isolation to achieve their objectives, they have to enrol other actors into their strategies. As such objectives need to be translated into action, to be made real, achieved through becoming embedded in relations of actors. It is from the specific strategies and tactics employed by actors – compromises, coercion, etc. – that individual and collective interests are seen to emerge. They are relationally produced. Translation aims therefore to capture the sense that interests and identities, as well as the relationships through which they are enabled, need to be negotiated. It is a process through which actors and objects are defined and brought into relation with each other through the creation of a new network of relations (Law 1999).

Callon (1986, 196) argues that translation provides a framework for the analysis of controversies: the way in which power is exercised as actors seek to impose themselves and their definition of the situation on actors around them. This is, of course, usually a competitive process as actors may resist the roles ascribed to them and the representation of the world. Translation does then refer to the process whereby actors are brought into a – however temporary – alignment of interests. Crucially, running concurrently to the formation of relations is a process of producing knowledge about the world (Callon 1986, 203): be that the identities and interests of the actors themselves or the best way in which government should be involved in water management. By exploring the interactions between actors, translation can then be seen as a means of exploring how neo-liberal representations of the world become authoritative.

A specific focus of ANT has been scientific controversies, in particular: "how and why do scientists construct contradictory versions of the 'facts'?; when there are opposing versions, how and why do some become 'true', while others fade into obscurity?" (Garrety 1997, 729). In *Science in Action*, Bruno Latour (1987) described science as having two-faces, or sides like Janus: "one that knows, the other that does not know yet" (7). The former, 'ready-made science', captured the way in which actors – wittingly and unwittingly – rationalised decision-making processes with hindsight. The latter, 'science in the making', referred to the contingent processes of translation through which science was produced. To get beyond the ready-made view of science, whilst understanding how it was constructed, Latour argued it is necessary to open-up the 'black-boxes' of science. This term, 'black-box', drawn from engineering, refers to "that which no longer needs to be reconsidered, those things whose contents have become a matter of indifference" (Callon and Latour 1981, 284). It refers to the crucial step through which once complicated processes with uncertain outcomes come to be seen as dependable; portrayed with certainty as outcomes become more reliable.

In this science in the making perspective, attempts at establishing order, of maintaining existing relations or of enrolling actors in a new network are constantly being resisted or re-defined according to the competing visions of actors. Change and stability in society emerge from the competition between actor-networks, who are, to varying degrees linked to a version of the 'truth', a view of the world, which provides a purpose for their action. These networks are built through the 'translation' of interests; possibly detaching actors from existing networks, enrolling them in new ones (Callon *et al.* 2002). The form the network takes is a result of the negotiations between all the enrolled actors and thus is not merely the outcome of the objectives of only one actor (Murdoch 2003, 271). Importantly, the process of translation aims to reveal how objectives are often mediated not merely by the negotiations which take place with humans but also by the performances of technologies and nature. Thus humans rarely act 'rationally' as all cognition and action is mediated by the social-technological-natural relations work within.

Actors who can construct the largest networks through allying themselves with the most actors – thereby translating the interests of others into their networks – are able to impose their view of the world on others, marginalising opposing actor-networks and visions (Callon & Latour, 1981: 292). To finally translate then, to mobilise an actor-network is to "express in one's own language what others say and want, why they act in the way they do and how they associate with each other" (Callon 1986, 223). Or to put it another way, "to speak for others is to first silence those in whose name we speak" (Callon 1986, 216).

In this view, power is some ways Foucauldian in the sense that it is earned and exercised and not simply possessed. "Power is not an institution, and not a structure; neither is it a certain strength we are endowed with; it is the name that one attributes to a complex strategical situation in a particular society" (Foucault 1998, 93). Power is achieved through networks and alliances of actors. Or as Callon (1986) states: "understanding what sociologists generally call power relationships means describing the ways in which actors are defined, associated, and simultaneously obliged to remain faithful to their alliances" (224). The concern in both approaches is not to assume that particular power relations exist but rather to expose the techniques and strategies by which power relations are achieved (Law 1991, 169). The governmentality literature is also informed by this Foucauldian view of power. As stated, however, the key difference lies in the treatment of discourse. The approach of Miller and Rose begins with the premise that discourses are constitutive of social relations, and traces how these discourses are produced and re-produced in different contexts and the effects these processes have. ANT, on the other hand, starts from the premise that discourses are an outcome of networks and explores the processes through which these networks are assembled. ANT writers have not, however, specifically studied the peculiarities of the realm of politics, and the appropriateness of this indeterminacy will be discussed below.

Through this process of translation, ANT aims to show how certain actors acquire power. Seeing power ultimately as an effect of networks, does not, however, mean that more conventional modes of power, such as those formally prescribed by political systems (the power to do something and the power over someone else) are entirely rejected. Rather actors that have such powers are seen as enjoying the effects of a stable network of relations in which power is stored (Law 1991, 166). Power is therefore displaced across a range of actors in the sense that the exercise of power is contingent on the performance of a network of actors. The loss of formal political powers would, from this perspective, be seen as the breakdown of the network of relations and/ or the successful contestation of the network by another with greater resources (a new form of governance, for example). Ultimately, in explaining the power and stability of networks in terms of the size, ANT combines a realist dimension (Jasanoff 2006, 23) with a more post-structural understanding of the fluidity of meanings and the instability of social structures.

In order for a network to endure across time and space, for power to be exercised at a distance, the network has to be constantly produced and reproduced. In other words, sets of relations, meanings, practices (such as privatisation) have to be translated and re-translated in very specific contexts of time and space (Mclean and Hassard 2004, 494). Translation is an ongoing process, one inherently susceptible to contestation. As a result, actors, objects and the meanings which bind them together in actor-networks are not 'fixed' in form and function but subject to a series of transformations (Latour 1999, 15).

Actor-network theory and policy analysis?

Having provided a brief overview of ANT, an obvious question would be, why utilise an approach from science and technology studies, concerned primarily with the analysis of innovation and scientific controversies, to study the policy-making process? After all, ANT is a perspective for understanding how entities – often non-human as well as human – are brought into relation with each other, how social structures take shape through the construction of actor-networks. First, it should be stressed that this study does not intend to adopt the "generalised symmetry" (Callon 1986, 200) of ANT and include technologies and natural entities in the analysis. Nor does it intend to get involved in the debates over human and non-human agency this has provoked amongst social scientists.[19] Instead, this study contends that

[19] See Bloor (1999) and Collins and Yearley (1992) for criticism; Hacking (1999a; 1999b) for a middle-ground perspective and Murdoch (2001; 2003) for an outline of how such a perspective could be applied to the study of nature-society relations.

policy-making is a realm in which human agency is generally to the fore and in the case of the BWB partial privatisation, it dominates and determines. It is primarily defined through institutions and the actions of people within or around them. This is not to say that the inclusion of natural and environmental entities in the analysis is always unwarranted in the study of policy. In his studies of governance in the EU and policy sectors such as genetics, Gottweiss explicitly draws on elements of ANT with regard to the treatment of non-human actors (1999; 2003). As interesting as Gottweiss's work is, his empirical focus is a field where science and policy-making explicitly overlap: where the researcher finds experimentation and instability concerning the ordering of the natural, technical and social. Generally, water privatisation debates do not display such instability of relations between human and non-human entities. Certainly in the BWB privatisation, facts about water, the infrastructural network, remained relatively ordered.

As stated, water governance has been bound-up in theories about the state and economic dynamics. Hence, approaches to water governance aiming for a more holistic appraisal of the economic, political and natural processes embodied in WSS still rest, first and foremost, on theories of the social – how human activity orders or attempts to order relations in water systems. An example would be the "political ecology" perspective (Castro *et al.* 2003; Swyngedouw *et al.* 2006) which stresses that the governance of water is a multi-dimensional activity. Within this perspective some writers have critically engaged with the ANT literature and, more broadly, the literature on human-non-human relations (for example Gandy 2005; Swyngedouw 2006).

A more holistic approach, such as that provided by political ecology brings much to the analysis of water and environmental governance. However, the focus of this study is more exclusively on the policy-making process through which the privatisation was realised. The aim is to provide an account of how privatisation was discursively and institutionally produced. Thus, the focus has to be on the peculiarities of the policy-making realm: the formal structures and processes of governance, the rationales guiding them and the philosophical debates underpinning them. Infrastructural and environmental concerns were largely absent from discussions. As such the inclusion of natural and environmental entities would add little to the analysis.

Policy-making as translation

Of particular relevance to this study of policy is the view that the ability to act is seen as being an outcome of relations, not an inherent quality of entities. Power is displaced across a range of actors. There is an emphasis in ANT on describing the ways in which objects and individuals often take on different qualities as they are translated into networks of relations (Rutland and Aylett 2008, 628); a focus on the structuring of power relationships which enable new practices and objects to emerge – the duality of knowledge production and structuring of power relations (Callon 1986, 203).

Thus with its sense of the fluidity of meaning and structures, of impermanence and almost Foucauldian conception of power as being displaced across a range of actors, ANT seems to strike at the heart of what commentators mean when they talk of governance. 'Governance' points to a looser, more spontaneous, dynamic political system, characterised by ideas of self-organisation, networks and partnerships which displace power across both public and private actors. Governance attempts to capture the sense that hierarchical and fixed structures in politics have become less apparent, as emergent forms of structure and decision-making have increased.

Translation is a useful means of analysing the new, unstable political spaces (Hajer 2003b) which characterise shifts to governance. It links the production of new knowledge with the creation of new networks of relations between actors. It thus captures the sense that the institutional forms politics takes can, at least in part, be produced through the making of policy. Policy and polity are co-produced when the formal rules and norms (Hajer 2003b) are not clear. Having first been applied to the study of scientific controversies and innovation, translation is an apt way of describing the processes through which rules and norms have to be agreed in unstable political systems. Furthermore, the ANT view of power as an outcome of networked relations, displaced across a range of actors, provides a more fitting way of thinking about political power in the context of a decline of formal political power and hierarchical decision-making.

With its emphasis on instability – of the transformation of interests and identities, meanings and objects – ANT seems better equipped to deal with the empirical object of 'governance' than traditional approaches. In particular it offers an alternative to the dominant 'policy networks' approach to policy studies, which emerged from a desire to capture policy-making beyond the traditional institutions of the state (Richardson 2000, 1006). Most commonly associated with Rhodes (1990) and others (see Marsh and Rhodes 1992), the weakness of the literature is its inability to characterise policy change, to conceptualise dynamism in policy-making. The notion of the policy network is founded on the "implication of stable policies, as well as stable relationships and a stable membership" (Richardson 2000, 1007). Policy-making is usually a more unpredictable phenomenon – less controllable than Marsh and Rhodes (1992) and others have suggested (Richardson 2000, 1007) with their "static" networks (John 2003, 486).[20]

[20] A criticism Rhodes (2006, 437) has accepted.

By contrast, an ANT perspective encourages us to see the identities and interests of actors and the networks they are embedded within as being fixed only through interaction, with other actors. In other words, interests are not fixed in advance and cannot be seen as a cause of action – they are defined through action and in the process power is effected. Thus though conventional forms of power exist in policy processes (the power over someone else, or the power to do something), these exist only in the extent to which they are performed, or effected by the networks of relations which political systems embody.

Finally, this approach to policy-making emphasises the dual movement of (re)-representation and (re)-association. It highlights the potential for "change, adaptation, mutation, and transformation" (Freeman 2007, 12) in policy-making in a context of governance. The implications of such a view of policy-making are that interests and identities, roles and responsibilities, are fundamentally negotiable. Translation is therefore a powerful concept in that it captures the dynamism of processes of construction through which policy is achieved: the contingency of meaning, the contests over facts and the construction of actors' roles and relationships.

This study does not intend to perform a whole utilisation ANT, though the potential may be there for a more detailed application of, for example, Callon's (1986) methodological framework of the sociology of translation. Such an approach, with its roots in ethnomethodology (Latour 1999, 19) requires a detail in observation which this study, with its analysis of a privatisation case from 10 years ago, was not able to provide. Instead, as the next section outlines, the study, argues that the notion of translation does provide a useful way of thinking about how political discourses become real – through action – in particular contexts. Further, the methodological and epistemological stance of the literature provides a useful corrective to the discourse-heavy, top-down tendencies of the governmentality literature. Of particular importance is the instruction that researchers should allow actors their voice because "actors know what they do and we have to learn from them not only what they do, but how and why they do it" (Latour 1999, 19). With this change in emphasis a tension emerges in the approach between discourse and agency, one which the study discusses but leaves unresolved.

The governmentality perspective and the notion of translation

Miller and Rose view the interplay between the macro (political rationalities) and the micro ('technologies of government') levels of government as a process of translation. The mobilisation of technologies of government can be seen, in ANT terms, as the translation of actor-networks: the means through which political programmes such as privatisation are rolled out in a particular field. They are the "strategies, techniques, and procedures" through which actors attempt to make their programmes operable (Rose and Miller 1992, 183). However, Miller and Rose do not delve that deeply into the contingency of translating political discourses or programmes such as neo-liberalism. In contrast, following an ANT perspective, contingency, the processes through which networks of actors are constructed, becomes the object of study. It locates the analysis in the local contexts of politics through which political practices are actually made and takes these interactions as its empirical focus, tracing their course to discover how new relations of power and bodies of knowledge are achieved as an outcome of these interactions.

Both ANT and governmentality can be said to have a concern with the practical aspects of governing, arguing that political discourses only have effects in the extent to which they become technical (Larner and Le Heron 2004, 213). Even keeping the 'non-human' issue out of the discussion, there is a tension between the two approaches regarding the means through which they seek to reveal this. An ANT perspective takes the view that there is a material world, something of tangible substance, but that it is defined inter-subjectively through the heterogeneous actor-networks of which it is constituted (Law and Singleton 2000, 2). Law (1994) has described it as "relational materialism". At its core, ANT views reality as being constructed through the interaction of a number of actors and this reality exists outside of the minds of any individual. The assumption that reality consists of actor-networks, or more precisely the effects of actor-networks, moves the researcher from a focus on actors, to the interactions between actors. It is in this sense relativist. Indeed, actors are defined, made real, only in these interactions with others. One consequence of such a view is to reject *a priori* the existence of social structures – institutions, categories – as influences on behaviour. In other words, there are no social forces hidden behind social interactions, institutions such as class only exist in the extent to which they are performed (see Latour 2005).

While acknowledging the similarity of their view of power to that of Foucault's (Law 1991, 169), key ANT writers such as Law have nonetheless been critical of the sense that discourses almost produce themselves in Foucault's work (see Law 1994, 107). More generally, Latour is critical of interpretative approaches in the social sciences which seek to show how actors are somehow "manipulated by forces exterior to themselves" and require social scientists to "provide a theory of the social or even worse an explanation of what makes society exert pressure on actors" (Latour, 1999, 19-20). From an ANT perspective, neo-liberalism may exist but it does so only as an effect of actor-networks. Thus ANT argues against both the realist notion that there is a reality which exists beyond human intervention but also those social constructivist perspectives that suggest reality somehow exists beyond the comprehension of those involved in constructing it. ANT would then be sceptical of Dean's (2008, 16) suggestion that actors may not always be aware of the mentalities of governing to which they contribute.

Other researchers have also argued that governmentality studies pay very little attention to political agency (Rutland and Aylett 2008, 631) and the ways in which governmental practices are built from the bottom up (O'Malley *et al.* 1997). A Foucauldian discourse approach shows "how a piece of given music comes to be performed by multiple actors, but tells us little about how the piece was chosen or composed" (Rutland and Aylett 2008, 631). The governmentality perspective, for instance, focuses its analytical gaze first on rationalities and then – how these constitute – actions. In contrast, ANT looks first to actions to describe how meanings and relations between actors are produced: it is 'bottom-up', rather than 'top-down'. The emphasis is on how discourses take their material form through negotiations between actors. In such an approach the actors themselves come to the fore in the analysis. They make their own worlds as Latour (1999) states.

Such an approach certainly has much to offer in methodological terms, particularly in relation to the study of exchanges between actors in policy-making processes. The emphasis is thus placed on description first and analysis second. In many ways, ANT is not a theory but primarily a methodology, as some of its key proponents have themselves stated (Latour 1999, 22; Callon 1999, 182). However, this absence of a theory of the social is problematic, particularly for the study of politics. In ANT there is little sense of the moral or political arguments which are usually inherent to the formation of systems of governance (Jasanoff 2006, 23). One consequence of this for a study of policy-making is that the exercise of power loses its specifically 'political' quality. ANT has little to say about the ways in which "beliefs, values and ideologies stabilise some representations of the world at the expense of others" (Jasanoff 2006, 23). Thus, however useful the notion of translation is in emphasising the inherent links between producing new knowledge and the formation of new relationships of actors, it has to be adapted to the study of politics. There has to be a sense that something broader informs processes of translation and that everything is produced through them. For this reason a synthesis with the governmentality perspective is appropriate: the context of governing is shaped by the contests between political rationalities. As Dean (2008, 9) states, from such a perspective, an analysis of governing rests fundamentally on the 'political'.

The political concerns a particular relationship of power. Following Foucault, a power relationship is characterised by Dean as the strategic attempts by actors to influence the actions of each other. Power is a struggle, or a "duel" (Dean 2008, 9) between actors, defined by instability and regular resistances. Power is therefore reciprocal in the sense that it is an outcome of shifting relations between actors – it is a product of the relationship, not a possession of the actors. This distinguishes it from "domination" in which relationships are relatively hierarchical and stable (Dean 2008, 9). The project of governing is bound-up in relationships of power as it rests on struggles to act on the actions of others. Governing societies:

> "deploys ways of thinking, or rationalities, and ways of intervening, or technologies, to try to order, fix, stabilize or even disorder and reverse relationships of power and bring about the downfall of states of domination" (Dean 2007, 9-10).

Neo-liberalism can be seen as such an attempt to problematise and re-configure state-society relations. It provokes a conflict over the role of government and markets in governing society. To study the political is thus to analyse the conflicting struggles, or duels between actors, through which political power is exercised. Translation may be more useful than the governmentality approach for exploring how local strategies emerge and shape these duels, but what it lacks is the sense of rationalities which inform them: the interplay between broader discourse and political action. In this sense, although there are tensions between the two approaches, combining them allows for a fuller analysis of processes of making policy, whilst addressing some of the weaknesses inherent to each. Thus the aim in this study is not to seek a resolution of these tensions but rather an accommodation – however uneasy – between them. The tensions should, further, be seen as productive for the analysis.

Conclusion: summarising the analytical approach

This chapter has drawn together the key insights from the literatures discussed in this and the previous chapter (post-positivist policy studies, the governmentality literature, the policy discourses and translation) and summarises how the BWB partial privatisation and policy-making in general will be seen. Overall, then, from this perspective policy-making and policy implementation can be seen as an attempt to define, delimit and design a discrete sphere of activity embodying a mentality/ mentalities of governing. Policy is as much about representing an aspect of the world in a particular way as it is a means of establishing governmental practices. If neo-liberalism defines a discursive field in which power is rationalised (Lemke 2001, 1), privatisation is about re-programming state-society relations in a defined realm. Whilst making claims, producing knowledge about the world, policy-making is simultaneously about enrolling actors in the creation of new practices to govern it.

Political rationalities and political programmes provide shape to policy-making. Political programmes in particular can be seen to provide blueprints for how to re-configure governmental practices. Privatisation brings with it a certain notion of what actors should do; it implies a (re-)representation and (re-)association of actors within a broader problematisation of state-society relations. Policy-making from this perspective captures the sense that policy – both in its formulation and implementation – is about the generation of a "collective script", which serves to define a "collective problem, a solution and a course of action" (Freeman 2007, 2). Through processes of (re-)representation and (re-)association, it "gives actors a sense of what they and others are or should be doing" and as such it "coordinates actions, ordering and often re-ordering (defining) a field" (Freeman 2007, 2). In this sense, "policy moves", motivating actors to carry out specific tasks and to accept certain responsibilities (Freeman 2007, 2).

Policy-making as translation does then rest on the simultaneous production of knowledge and the construction of power relations (Callon 1986, 203): the attempt to effect identities, interests – to constitute actors – as well as outcomes and practices. Rationalities, policy discourses and political programmes inform this process whether in the form of storylines, policy vocabularies or generative metaphors. The 'global city' is a good example in relation to the study of 1990s Berlin. Policy-making is not, however, 'determined' or 'derivable' from such rationalities, narratives or symbols (Miller and Rose 2008, 61). They can inform and inspire policy-making but can just as easily become transformed in processes of translation – come to mean different things in different policy sectors at different times.

Rationalities and the typical programmes of government with which they are associated, play through, are revised and sometimes rejected as policy is made within local contexts. It is important to remember that this policy process should be seen as the interplay between macro and micro processes. In Berlin, for example, there are the unique historical conditions produced by the Cold War: the isolation and subsidisation of the West, the socialist government of the East and the vast array of challenges – and opportunities – provided by bringing the two halves together and re-integrating the city in the global economy. These characteristics are entirely peculiar to Berlin in this period as are the CDU-SPD coalition politics. The form rationalities and programmes take is contingent upon an array of contextual factors particular to a place and time. Thinking back to the notion of governmentality, it is in this sense, that there is never simply an 'application' of neo-liberal rationalities, but only actually existing neo-liberalism.

More fundamentally, policies, with all their compromises and local contingencies, can be said to embody mentalities of governing. They create both discursive and material spaces (practices of government) in which a particular view of governing is rationalised. The making of a policy contributes to the translation of a mentality of governing: the sense that one mode of thinking – from a range – can come to be seen as rational, taken for granted. As a mentality of governing is embedded, the ways in which norms and theories about the world (political rationalities) shape the production of 'facts', becomes less apparent. It can obscure the normative and intellectual means through which one representation of reality is presented as the only or superior version of reality.

It is this context that the 'no alternative' to privatisation argument emerges and becomes potent. Chapter 6 outlines this 'no alternative' argument, placing it within an assessment of the formal policy-making process. In doing so, the 'no alternative' account of why the privatisation occurred is shown to merge with official versions of how the privatisation was actually carried out: largely through the formal institutions and decision-making processes of the Berlin political system. Taken together, these accounts are critiqued as being ready-made (Latour 1987): rationalisations of the process which simplify and reduce contingency and controversy, obscuring the biases on which the account rests with claims to rationality.

A distinction is made between ready-made politics and politics in the making accounts of the policy process. Most pertinently, how many participants 'black-boxed' the policy process in the 'objectivism' and 'rationality' of the formal institutional processes. In the BWB privatisation process this refers to the concealment of conflicts over policy strategies (should Berlin really aim to become a global city?), debates over facts (just who and what was to blame for Berlin's debt?) and conflicts over the identities of objects in policy-making (should BWB be re-defined as a commercial player?). A key task of this study is to show how it became possible to present privatisation in straightforward terms even in the still relatively 'leftist' political context of Berlin. How was the 'no alternative' argument constructed? The following chapters will set about opening-up the 'black boxes' of the BWB privatisation, revealing some of the steps through which the relationships and knowledge were established to translate the privatisation.

These ready-made versions, provided by civil servants, politicians and the hired consultants, allude to the formal institutional processes of the political system and the legitimacy and objectivity that they claim to provide. Clearly institutions remain integral to the making of policy but from the perspective adopted in this study they are not seen as having stable functions which determine the practices of policy-making and implementation. Of course, policy-making does not occur on a blank canvas – the terrain of governing must be accounted for. Political rationalities inform the making of policy. They also suggest ideas as to how the political system should function: what the role of political institutions should be, how societal actors should be included in governance structures. Although the roles and responsibilities of institutions are formally designated by political systems, it is argued that they should not necessarily be seen as fixed, but open to negotiation. Instead, they must be achieved – agreed upon through the policy-making process itself. From such a perspective, actors, institutions and the logics which inform their interaction are bound-up in translation. The political system is as much an outcome of policy as it is an input (Hansen and Sorenson 2005, 93). The roles of actors and institutions and their relations to each other have to be produced in policy-making processes.

The policy-making process is then constitutive of the political decision-making system (Hajer 2003b). A political system, regardless of the theories and traditions which give it a certain coherence and constancy, must still be translated. It is constantly being achieved through policy-making and thus its exact form, the configuration of relations between institutions, political parties and societal actors can change. The exercise of power may be formalised in the institutions of the political system – executives have the capacity to initiate legislation, to work, manipulate the decision-making system, to coerce their parties, and so on. However, the exercise of power is never as simple as that. Formal political power may act as a 'push' but it can never fully explain the process of exercising power. Power must be effected; it is the achievement, or outcome of action, not merely the cause of action. In this respect the policy-making process is constitutive of power. Furthermore, participants in the policy process or the objects of policy-making are participants in the exercise of power. They may not all benefit from the power relations established but they are shaped by them: roles and responsibilities are assigned, interests and identities ascribed. It is in this way that Foucault (1991; 1998) saw power to be 'productive' or 'positive': not in any moral sense, but in a material sense.

In analysing the BWB partial privatisation, the study explores the relationships between neo-liberalism and the emergence of new forms of governance. It provides an account of how the political system functioned; the distribution of roles and responsibilities in decision-making; and how political power was exercised. The case study analysis also aims to reveal something of the nature of new forms of governance in the paradigm of neo-liberalism by highlighting the roles played by consultants in the process. It is argued that consultants – hired private sector economic, financial and legal consultants – became important political actors. Private sector consultants could be found lobbying for privatisation, proposing complex models of privatisation and managing the bidding process for the Finance Senate on behalf of the Berlin government. They were appointed to these roles because the Berlin government felt it required the knowledge of privatisation deals and, more broadly, their experience of working with large private companies.

The consultants who managed the bidding process provide a particular focus. It will be shown that they were not merely consulting politicians, but practicing politics. They took on many duties which would previously have been performed by elected politicians or civil servants (who, at least theoretically, are bound by notions of public service and the need to represent their office's/ department's interests). Of course, experts of one form or another have always been involved in political decision-making. What is new, however, are the specific roles these consultants are playing, their relationships with the political actors with whom they work and the means through which their roles are produced through neo-liberal programmes such as privatisation.

By bringing together the sociology of translation with governmentality and interpretivist approaches to policy studies, political rationalities are seen as informing actions and, in turn, being informed by actions, as they are reformed, rejected and realised in very particular contexts. This synthesis of the approaches aims to, if not capture this interplay, then expose both sides of the process. Of course the tension remains unresolved here. On the one hand, the approach points to the effects of discourses, the way in which they shape agency – what is seen as possible and what is not. On the other, there is the sense that discourses are but the outcome of the actual interactions between actors. As such there are two dimensions to the analysis.

As stated, Chapter 6 focuses the analysis on the formal period of the policy process, examining how the institutions of the political system actually functioned. Prior to this, however, Chapter 5 assesses the broader policy-making context in 1990s Berlin. It argues that it is possible to understand some of the key political decisions in the decade as a series of translations of the global city policy discourse. It was a process through which the very particular conditions in Berlin came to be processed using the moral and intellectual machinery of neo-liberalism (Lemke 2001, 2). Through this a Berlin-specific version of neo-liberalism emerged, one productive of failures, tensions and paradoxes. Ultimately, these translations shape the material and discursive context of the partial privatisation of BWB. They inform a politics of inevitability.

5. The global city policy discourse and water policy-making: making the privatisation of BWB 'inevitable'

Introduction

In January 1999, the Berlin government, in partnership with Deutsche Telekom, unveiled three new *Mediapolis* projects. These were presented as part of *Die Wende* ("the transition") from an industrial-based economy to an information society. The emphasis was on "innovation", "new media" and creating "knowledge networks" (Der Tagesspiegel 1999d). The high-tech, knowledge-rich sectors were the future for Berlin. This was the modish language of the time, the vocabulary of the global city policy discourse.

A month later, a reporter in the same Berlin newspaper, Der Tagesspiegel (Frese 1999) surveyed the latest reports on Berlin's financial and economic situation. They made grim reading, with the only positive aspect of the reports being their clarity, the paper remarked. This was the fifth year of rising governmental debts and negative news about Berlin's economy. Industry was disappearing, the restructuring of the economy stalling, while innovation actually dropped in some areas. Indeed, the optimism and grand plans which had characterised the decade should now be forgotten, the paper argued. The Olympic bid had failed, the proposed fusion of Berlin-Brandenburg into a city-region had been resisted by Brandenburgers and only the arrival of the Federal Government could work as an '*Impulsgeber*' ("a source of impetus") for the economy. At least, the paper continued, the news should, once and for all, bring an end to the illusions of "boomtown Berlin" and the *Dienstleistungsmetropole* ("Service Metropolis"). More accurately, they concluded Berlin should be called the *Arbeitslosenmetropole* ("Unemployment Metropolis"). Concluding, the paper stated, 'Berlin braucht viel Geduld' ("Berlin needs a lot of patience").

A week or so later, the Economy Senator, Wolfgang Branoner (CDU), struck a more optimistic note: "Truly, we will become the most 'state-of-the-art' city in the Western World" (Böhm and Hasse 1999). The message was clear. Things may indeed be bad, but it is just a matter of time before Berlin will prosper. The emphasis in this period – and still to a lesser extent today – was that Berlin was *becoming*: a 'global city', a *Dienstleistungsmetropole*, a 'gateway city' to Central and Eastern European markets.

These articles capture the mood of the period in which BWB was privatised and something of the thinking and practices of policy-making. On the one hand, the rising disenchantment in the city as the economic situation deteriorated; the continuing grand plans of the Berlin government, on the other. There was a cyclical dynamic to policy-making. New policies aimed at re-structuring the economy were announced with great rhetorical flourish. Often such policies had been devised and were to be realised in partnership with the private sector. It would soon, however, become apparent that Berlin's general economic and financial situation had further worsened. Rather than leading to a wholesale re-appraisal of policy, yet another new policy aiming at developing 'knowledge' sectors of the 'new media' in Berlin would be announced. This is not to suggest that such policies were always the cause of the worsening situation in Berlin. Rather it is to state that these policies did not make the situation much better.

It was within this context that the partial privatisation of BWB was proposed. The aims were two-fold: to raise finances to ease Berlin's fiscal troubles and make BWB "competitive" in international water markets, as the Economy Senator Branoner stated upon completion of the privatisation deal in October 1999 (Fischer 1999). This chapter explores the reasons for this decision. In doing so it moves beyond the simplified accounts usually given for the privatisation: that the Berlin government needed the money and privatisation was the logical policy option available. The chapter takes issue with the claimed objectivity and simplicity of this version. It can be seen as an official, ready-made politics account of the BWB privatisation. The Berlin government was struggling to cope with the costs of reunification and in the political climate of neo-liberalism the response was to sell off state assets to reduce the running costs of Berlin and alleviate debt. Here the decision to privatise is necessary because there was no alternative: policy-making becomes a simple process of needs must.

This is shown to be one representation of the decision-making process. It is a construct which, while making claims to neutrality and objectivity, conceals neo-liberal assumptions about policy-making and the contingency of the decision upon a range of prior policy decisions. The study accepts that, by the time the privatisation of BWB emerged as a policy option, it was indeed possible to argue for it in straightforward terms. Thus, though this chapter seeks to move beyond the ready-made account of the BWB privatisation, it also aims to understand how an inevitability emerged, suggesting how it was constructed. To do this a different understanding of the interplay between discourse, socio-economic dynamics and the practices of politics is required. It aims to show that the privatisation of BWB was bound up with the aspiration to become a global city, the policy means employed to achieve it, and the effects these means had on BWB, the Berlin government's finances and the socio-economic situation within the city.

This alternative account, or politics in the making account, centres on the potency of the global city policy discourse as discussed in Chapter 4. This policy discourse captured the way in which the government and its advisers brought the global and local together: the way in which Berlin's re-entry into a globalising economy was interpreted. It represented a neo-liberal interpretation of the tasks facing the newly reunified city, defining 'problems' and outlining solutions. The policy discourse did then bring the New Berlin into relation with the global economy.

The chapter is structured as follows. First, a ready-made explanation of the privatisation of BWB provided by a civil servant working at the Finance Senate is presented. This account is then problematised in the second section, which discusses policy-making and the socio-economic context of 1990s Berlin from the perspective of the global city policy discourse. In this politics in the making account, emblematic translations of the policy discourse are discussed, including that of Bankgesellschaft Berlin. The gap between the symbol and realities of global city policy-making is made clear and the problems inherent to this policy-making agenda in Berlin are emphasised. The third section moves onto the translation of the policy discourse in water policy-making, following the attempts to make BWB a 'global player' and the further problematisation of BWB in the context of the apparent failings of BWB and Berlin as both encountered fiscal problems. Next, the privatisation policy process, from formal proposal to formal approval by the government and Parliament (June 1997-July 1998), is analysed. The 'inevitability' of the proposal is discussed in the context of the negotiations required to make it a policy. It is argued that there was indeed an inevitability to the privatisation but that it was produced. This can best be explained with reference to the material and discursive effects of global city policy-making, on the one hand, and by looking in detail at the negotiations between actors in the policy debate, on the other. The prior translations of the discourse mobilised bias and shaped the terms of the policy debate, but making the decision still required work: the building of support, the isolation of opponents and so on.

A ready-made account of why BWB was partially privatised

In the ready-made politics version the privatisation is relatively free of controversy. Indeed, everything seems simple with hindsight. The civil servant at Finance draws a distinction between the official or public reasons given for privatisation and the real, underlying reason. The official explanation given to the media and, in turn, to the people of Berlin was that they needed private sector expertise to run BWB more effectively. Furthermore, they specifically needed international companies and the expertise they could bring and not merely that provided by German companies, be they multi-national or not (Interview 14). The civil servant stated that this was Fugmann-Heessing's explicit aim and can be seen as an example of the desire she and others had to bring international companies to Berlin (Interview 14).

Clarifying this objective, the civil servant stressed the importance of the time and touched upon the broader discourses about the state's role (Interview 14). In the mid-1990s the logics and practices of the private sector were seen very much as the answer to problems of governing, the civil servant noted. The state's function had changed: it had to become a regulator not an owner. It was, the civil servant added, not long after the Premiership of Margaret Thatcher and privatisation was not seen in such a critical light. Hence why there was, at least according to the civil servant, a consensus in favour of privatisation at the top of both political parties (Interview 14).

However, the privatisation of BWB was not, in the civil servant's view, driven by ideology (Interview 14). Rather, the 'real' explanation for the privatisation of BWB was the parlous state of Berlin's finances (Interview 14). Indeed, the civil servant stated that the rhetoric of privatisation (the efficiency of private sector ownership, etc.) was mainly for media and public consumption (Interview 14). The implication here is that the rhetoric was useful in justifying the policy, but it did not capture the 'real' reasons for the privatisation. Thus though the civil servant hints at the importance of the discursive climate of the 1990s, the real reason for the privatisation lay elsewhere: in an assessment of the hard 'facts' of Berlin's economic situation. Later, a different relationship between discourses, Berlin's economy and the actions of politicians will be drawn. At present, however, this ready-made account will be outlined.

"Berlin has always been poor, even before World War 2".[21]

At the time of interviewing the civil servant – shortly after the agreement of the partial privatisation of the *Berliner Sparkasse* (Berlin Savings Bank) in June 2006 – Berlin had, since reunification, made roughly $17.8 Billion (€14 billion) from privatisations (Interview 14). Despite this, the Berlin government was and still is struggling with huge debts. The civil servant confirmed that this has been partially self-inflicted through governmental malpractice, as exemplified by the 'Banking Scandal' which emerged in 2001 (described below) (Interview 14). Still, as the above quotation suggests, the civil servant saw the problems of the 1990s and the present

[21] Senior civil servant at Finance: Interview 14.

day as being more than just a result of the costs of reunification and fiscal governance (Interview 14). If privatisation was the direct result of the Berlin's government's grim financial situation, the grim financial situation was not the direct result of the Berlin's government's actions. The civil servant aimed to put distance between the two, providing historical context for the situation. Berlin's problems were long-term and structural; the city had never had a strong economy (Interview 14). The civil servant argued that such problems were, at least to some extent, concealed during the Cold War as both sides of Berlin enjoyed large subsidies, particularly West Berlin (Interview 14). Losing the money from Bonn upon reunification and the need for huge investment in East Berlin brought the (West) Berlin political class face to face with this reality.

Alongside the loss of subsidies and the need for investment in Berlin, the civil servant identified the duplication of personnel in the public sector as the key challenge facing the Berlin government in the 1990s, such was the financial burden on the state of having two employees (one from the East and one from the West) performing similar roles (Interview 14). In this account, the privatisation programme of the 1990s must be seen within the broader public sector reform programme implemented in Berlin by Fugmann-Heesing (Interview 14). Drawing on ideas of the New Public Management (NPM), Fugmann-Heesing wanted to implement a wholesale reform of the public sector. Efficiency measures were introduced in sectors such as education, the police service, the health sector and so on to reduce the government's outgoings (Interview 14).

The civil servant described this as a long and difficult process due to resistance from the Unions and the need to be sensitive to East Berliners. Although the civil servant stated that the reforms were not entirely successful, Berlin is still one of the few *Länder* (States) to have conducted such a wholesale implementation of NPM ideas (Schedler and Proeller 2002, 172). In line with the explanation given for the privatisation of BWB, the civil servant emphasised that these policies were not 'ideological' but born of necessity (Interview 14). The privatisation programme (Housing Associations, Gas, Electricity and Water companies) was, according to the civil servant, partly designed to buy time for this public service reform (Interview 14). In the short-term, Fugmann-Heesing was able to reduce the amount of money being borrowed by Berlin, if not the actual money that needed to be paid back (Interview 14). This account of the decision to privatise ended with an emphasis on the fact that Berlin had no real alternatives to privatisation: it had no real economy and never had: only small businesses and a strong cultural and scientific community (Interview 14).

A politics in the making account of the background to the BWB privatisation

To move beyond this ready-made account, and in the process provide a critique, the chapter now explores the potency of the global city policy discourse and the way in which it shaped policy-making. In the ready-made account there was no real discussion of the broader economic policy pursued by the Berlin government in the 1990s. Nor was there any real engagement with the challenges of re-entering the global economy after the Cold War. Instead, the economic situation was described by the civil servant with reference to the loss of subsidies. He described Berlin as a poor city, lacking a strong economic base. Indeed, he claims it was poor even before the upheaval and destruction of the Second World War. The financial crisis which emerged in the mid- to late-1990s was described with reference to the challenges and costs inherent to merging East and West Berlin. It was accepted that the government contributed to the financial situation apparent in Berlin in the 2000s (through the Banking Scandal) and that Fugmann-Heesing's reforms were not entirely successful in the 1990s. However, the government's economic policy in the 1990s, its pursuit of the global city aspiration, does not feature in this discussion of the financial and ideological context in which BWB was privatised.

The politics in the making account provides a different discussion of this context, one which views the BWB privatisation as being bound-up in the government's economic policy-making and the global city policy discourse. It is an account which places the global city aspiration to the fore. The Berlin government's continued hyping of Berlin can be criticised, but it must be stated that most – business, political, academic – experts thought Berlin was well-placed to grow into a key economic centre in the global economy (Gornig and Häusserman 2002, 334). Although employment was to drop steadily through much of the 1990s, at the start of the decade job growth of 200,000 had been forecast by some experts based in large part on an expected boom in service industries in Berlin (Gornig and Häusserman 2002, 334).

Given such developments, the following section provides a critique of the global city policy-making. What the global policy discourse neglected, or at least, underestimated was not so much the extent of the changes that had occurred in the global economy during Berlin's Cold War isolation, but rather the extent to which Berlin's economy and economic functions had changed during this period. Most importantly, the big cities of West Germany became key centres of industry and finance during this period, while Berlin lost many of its economic and cultural functions. These were real and structural disadvantages which Berlin had to overcome and ones which were not directly addressed in the global city discourse. The promise of the discourse was that Berlin had the attributes to become a global city, or that it would somehow simply regain the economic and cultural functions it had enjoyed prior to the Second World War. This would be achieved through economic restructuring, which would make Berlin competitive in the global economy. Thus the global city discourse directed attention to the apparent needs of the global economy while overlooking the local and national economic context. The continued pursuit of the global city aspiration, despite much evidence to suggest that the policies were not working, is ultimately not indicative of an objective assessment of the situation.

The legacy of Cold War isolation and the realities of becoming a global city

If the rhetoric of the global city discourse is already clear, from what context was it emerging – what were the economic and social conditions in Berlin at the start of the 1990s? A good point of departure is to contrast the image–making with the more concrete changes reunification involved for the residents of Berlin. The end of the Cold War may have brought optimism but in terms of daily life it also brought great upheaval. East Berlin lost its status as the capital city and economic centre of the German Democratic Republic. It had been the seat of government and the main centre of medicine, education and media (Ellger 1992, 43). The disbanding of the socialist state apparatus between 1989-1992 resulted in the loss of 40% of the jobs in East Berlin (Gornig and Häussermann 2002, 335).

In the West, the end of the Cold War brought to an end the generous subsidies of the 'island' economy. In the late 1980s, West Berlin received $11 billion (DM 20 billion) per year to bolster the city's budget, to subsidise industrial production in the city and to finance an 8% tax reduction for all inhabitants (Ellger 1992, 42). Such generosity was not to last in the 1990s. Federal Government grants to Berlin continued, albeit on a lesser scale, but were severely reduced in 1993 (Newman and Thornley 1996, 98). An important consequence of the Cold War period was that West Berlin's industrial manufacturing sector had existed apart from the wider capitalist economy, surviving only through governmental subsidies aimed at bolstering employment in West Berlin. Upon exposure to the global economy, without the generous subsidies of the past, much of these businesses disappeared. Equally well-supported, or "over-represented" (Ellger 1992, 43), during the Cold War was the public sector which would become one of the main targets of governmental reforms in 1990s.

The Berlin-based journalist stated in an interview that the politicians in Cold War West Berlin "had so much money they didn't know what to do with it" (Interview 4). Beyond the hyperbole of this statement, it is clear that living in this 'bubble', with a steady flow of financial support, left an imprint on the way in which politics were conducted in Berlin during the 1990s. It was in large part the politicians from the old West who took control of the city government in the 1990s (in a CDU-SPD coalition government, 1990-2001). Some interviewees stated that they brought with them a lack of financial prudence which was entirely unsuited to the challenges Berlin was facing as the costs of reunification grew and subsidies declined (Interviews: 2, 4, 12). Additionally, it was argued that a "clubby atmosphere" (Interview 4) continued in post-Cold War Berlin politics: an 'old boys' network' where the same old faces within the CDU and SPD held power and made decisions (Interviews: 2, 4).

An alternative storyline: structural obstacles to becoming a global city

The civil servant stated that Berlin has always been poor. The implication was that the Cold War isolation of the West and socialism in the East made the generally bad economic situation worse. It would be more accurate, however, to argue that the Cold War caused a steady decline of the city's economy in relation to the wider capitalist economy (of Germany and Europe). Prior to the Second World War, Berlin had experienced periods of economic growth. After the unification of Germany in 1871, Berlin became the capital and this new status, combined with industrialisation and a huge rise in the population brought economic growth and great wealth to some in the city (Ellger 1992, 40). Indeed, such was the growing importance of the city and the strength of its economy, that Berlin quickly overtook Frankfurt as the main financial centre in Germany after unification (Wilson 1986, 152).

The decline of Berlin's status within Germany during the Cold War was apparent in 1989. In 1939 Berlin's share of the total German workforce was 10.3%; in 1989 it was 5%, despite the subsidies to the manufacturing industry of the West and the East's status as the capital of the German Democratic Republic (Gornig and Häussermann 2002, 332). The flipside to the optimism and hype of the early 1990s was the end of subsidies, the collapse of manufacturing industry and a consequent rise in unemployment.

Thus, one of the main problems faced by the Berlin government during the 1990s was finding employment for those who had previously worked in the manufacturing sector (Ellger 1992, 45) and the public sector. As stated, policy-makers aimed to re-structure the economy and focus on service-based, high-tech, knowledge sectors, as found in prosperous European cities. The hope for Berlin was that there would be a boom in other sectors to compensate for the jobs lost. The loss of jobs in the industrial sector was partially offset in West Berlin by a short period of growth until 1992, but after this came to an end, unemployment rose, albeit to a less severe extent than in East Berlin (Gornig and Häussermann 2002, 335). Thus, despite a speculative property boom in the early and mid-1990s, the expected growth in Berlin's economy never happened.

Obviously, the challenges facing the Berlin government were great. The costs of reunification were high and adapting the city's economy to neo-liberal globalisation undoubtedly difficult. The study does not dispute this. Rather, it problematises the global city aspiration and the policies implemented to realise it. It argues that the neo-liberal discourse of inevitability centred on the global city aspiration obscured the obstacles to growth in many of the sectors Berlin was seeking to develop, whilst justifying increasing hardship.

The long-term effects of Berlin's isolation from the Western European economic system were not fully accounted for in the making of economic policy. Instead, the government implemented a range of neo-liberal polices commonly found elsewhere in this period. As stated in Chapters 1 and 2, neo-liberalism was embedded in the structures of the global economy. It also, however, had a discursive quality. It represented the global economy in such a way as to suggest there were no alternatives to neo-liberal globalisation. It shaped the apparent possibilities in policy-making. In this sense, policy-making was shaped by the norms of urban governance (Cochrane and Passmore 2001, 343); the belief that if Berlin conformed to the 'necessities' of economic globalisation it had the attributes to prosper. This is a tribute to the compelling quality of neo-liberalism – in the form of a global city discourse – in the period. Nonetheless, the lack of economic growth, high unemployment and the fiscal crisis which was to emerge must also be attributed to the particular policies and practices of the Berlin government and its advisers.

One of the most consistent critics of the global city vision and the Berlin government's economic policy has been the German geographer Stefan Krätke (2001, 2004a, 2004b). For him the problems which emerged in the city as a result of the speculative real estate boom are proof of the misplaced optimism of the global city vision and the incompetence of policy-making by the government during the decade (2004a; 2004b). Importantly, the failings of governmental policy and the economic hardships this was bringing were apparent by the mid-1990s but still the policies continued.

Krätke's main argument is that Berlin did not have the economic attributes to establish such a service-based economy. This is an important criticism, one which goes to the heart of government policy during the 1990s. As problems grew during this period they were framed within the storyline that Berlin was becoming a global city and that they were merely a necessary hardship along the way. In other words, the objective of becoming a global city, grounded in misjudged optimism and promoted despite evidence of an end to the post-reunification boom from the 1990s, came to justify other hardships. Ultimately, the privatisation of public companies became enveloped in this thinking, with Fugmann-Heesing apparently seeing privatisation as a way of "buying time" for Berlin to restructure its economy (Interview 7) and reform the public sector (Interview 14).

Competition with other German cities

After the Second World War, the occupying Allied powers explicitly sought to decentralise West Germany through encouraging the development of prosperous city-regions: Frankfurt-Main, Munich, Hamburg, Stuttgart and the Rhineland. Thus despite the return of capital status and the Federal government to Berlin in the 1990s, 40 years of isolation during the Cold War had allowed other cities in Germany to develop economic capacities enjoyed by capital cities in other countries (London in the UK and Paris in France, for instance). Prior to the Second World War, Berlin was an industrial base, the centre for banks, insurance companies, publishing, film, radio and television in Germany (Gornig and Häussermann 2002, 332-3). After the Second World War and the division of Germany, the major banks moved to Frankfurt; Hamburg and Munich became centres of the media industries and Stuttgart and Munich grew as industrial centres to complement the Ruhr-Rhine area (Gornig and Häussermann 2002, 332-3).

One of the key components of the global city vision was the transformation of Berlin into a service metropolis, providing knowledge-intensive services and goods. Krätke (2004a, 59) states that "new islands of economic growth" emerged in the software industry, bio-technology and medical engineering. There were phases of growth in the service sector, particularly in the business-related sectors such as financial and consulting services in the years immediately after the wall came down. However, as stated, after this initial boom up until the mid-1990s, Berlin came "into normal competition with other service industry centres", particularly those in Western Germany (Gornig and Häussermann 2002, 338). Here, the legacy of cold war isolation has proved difficult to change.

Problematically for Berlin, cities such as Munich and Stuttgart were already established as advanced centres of research and development in, for example, micro-electronics and bio-technology (Ellger 1992, 45). Furthermore, they were not simply centres within Germany but also within Europe. Berlin's aim to be globally 'competitive' was then hindered by its inability to overtake rivals within the German context.

Comparisons with other German cities and regions reveal that Berlin made little progress in catching up during the 1990s and that it was no closer to being the pre-eminent economic centre even within Germany. For example, Berlin was, in 2002, still well behind all other cities and regions (e.g. Hamburg, Cologne/ Bonn, Frankfurt-Main, and Munich) in terms of employment provided by credit agencies, insurance companies, legal and economic consultants, advertising and media companies (Gornig and Häussermann 2002, 339). Though some – at least at first – optimistically saw such figures as a sign of Berlin's potential for growth, Gornig and Häussermann (2002, 339) stressed that these findings were evidence that Berlin would not catch up with the better established economic centres of Germany. Such figures emphasise the concentration of businesses and expertise in these urban regions, the bonds between businesses and their locations. Indeed, having a concentration of businesses in a particular sector tends to lead to further consolidation, as the decision to locate the European Central Bank in Frankfurt-am-Main, the home of Germany's four largest banks, suggests.

As a result, by the early 2000s, there were only "islands of economic growth" (Krätke 2004a, 2004b). Berlin's particular strengths lay in its research and academic communities and its cultural and media sectors (Eckardt 2004, 3-5). The media and cultural industries, traditionally strong, grew and Berlin can be seen as a world-ranking 'Media City' (Krätke 2004a, 59). However, in an assessment of the two aspects of Berlin's economy usually seen as integral to global cities (supra-regional control capacity and the concentration of high-level business services), Krätke (2001) concluded that Berlin had virtually no chance of becoming a rival to London and Paris. Thus, by the end of the decade, government policy had failed to transform Berlin's economy; to replace the disappearing industrial sector with a service sector of supra-regional importance. A broader assessment of Berlin's service sector indicates that growth was strongest in low-qualification sectors. Berlin "might actually be characterised as the capital of cleaning squads and private security firms" (Krätke 2004a, 59).

Therefore, although Berlin was, like London and Paris, a multicultural capital and the seat of national government, its economic control capacities were "minimal" (Krätke 2001, 1782). Furthermore, there was "no realistic prospect" of Berlin developing into a global city (Krätke 2001, 1785). In fact, along with its strength in media and culture, the only other key resemblance Berlin has to a global city is in terms of the growing divides between rich and poor which have become apparent since reunification. With unemployment almost doubling between 1991-2003 (from 10% to 19%) and those claiming welfare benefits as a share of the population also doubling in this period, Berlin acquired "the socio-spatial divisions of a global city without attaining the economic power of a global city" (Krätke 2004a, 61).

Key economic decision makers in the German and global economy are still absent from Berlin, as big business has largely stayed away (Gornig and Häusserman (2002, 339). Most companies which had based their HQs in Berlin, for example Siemens and AEG, had by the 1950s left West Berlin on the grounds that it was not safe enough (Ellger 1992, 43). By the end of the 1990s Berlin had none of the knowledge-intensive branches of large corporations such as management, marketing and research (Ellger 1992, 43) seen as integral to a global city. Though some major companies, such as Sony and Siemens established HQs in Berlin, they were of second rank importance, covering only part of their operations (Krätke 2001, 1782). The early hope of establishing Berlin as a location for HQs had thus, by the end of the decade, "ended in resignation" (Eckardt 2004, 5).

Apart from the problems presented by competition with other German cities and regions, the idea that Berlin could become a 'springboard' or 'gateway' to the East also failed to materialise. In fact, much of Berlin's economy is now oriented towards the West (Krätke 2004, 512). In the end, Berlin's geographical position has proved to be a weakness rather than strength. It is isolated from the main centres of European economic activity, lying outside the so-called "blue banana" between London and Milan (Ward 2004, 243-4). This situation was not helped by the limited rail and air infrastructure connecting Berlin to other key centres in the European and global economy. Despite the construction of a huge new railway station, Berlin is still – after over a decade of discussions and plans – in the process of building an international airport. For much of the 1990s and early 2000s, Berlin did not have direct flights to key global cities such as New York and Tokyo, a fact which cannot have encouraged global players to move to the city and may have threatened the continuing presence of the few the city has (Ward 2004, 243).

Such were the problems in Berlin that its economic status within Germany actually declined in the 1990s, with its share of the German workforce dropping from 5% in 1989 to 4.3% in 1997 (Gornig and Häussermann 2002, 336). Additionally, the city's population levels began to decline during the 1990s. In the early 1990s predictions were made that the city's population would grow from 3.2 to 4 million by the end of the 1990s (Brügel 1993, 156). Though the population did initially grow rapidly – mirroring the mini-boom in Berlin – to just under 3.5 million in 1993, it then dropped every year to less than 3.4 million in 1999 (Amt für Statistik Berlin-Brandenburg 2007). If the optimism of most academics had begun to fade in the mid-1990s, then by the new millennium, it had virtually disappeared. It was argued Berlin could not even become the main economic centre in Germany, let alone a global city. "Berlin cannot and will not be the pre-eminent centre of supra-regional services that it was before the Second World War. Thus, Berlin will not again become the metropolis of Germany" (Gornig and Häussermann 2002, 340).

The persistence of the global policy discourse and global city policy-making

This is a damning verdict on Berlin's global city pretensions and the policies implemented during the 1990s to achieve it. Even the Mayor was forced to admit in 2000 that "Berlin will not succeed in reaching the status of a Global City in the next 10-15 years" (quoted in Eckardt 2004, 3). It also, of course, undermines the narrative of inevitable restructuring towards the service sector. Though this was already becoming apparent in the mid-1990s, the global city discourse continued in the face of growing scepticism. This optimism somehow managed to survive the unforgiving actualities of the 1990s, which exposed the structural weaknesses of Berlin's economy and its status within the new Germany and Europe.

The discourses of urban governance in the 1990s, informed by the perceived needs of globalisation and the means of neo-liberalism, may have been dominant but, as the last chapter made clear, ultimately polices are made and discourses produced by actors. The government of Berlin was perhaps Berlin's greatest booster and nowhere was this more apparent than in the commercial property market and the re-development of Berlin. Through their continued funding of a public-private financial corporation, Bankgesellschaft Berlin, they financially assisted the real estate boom. This material support was combined with rhetoric about Berlin's potential for growth. In the early 1990s, the government stated that 11 million square metres of new office space were required by 2010 to equip Berlin as a major service metropolis (Krätke 2004a, 62). By the end of the decade Berlin had the largest unoccupied office spaces in the world: 1.5 million square metres (Krätke 2004a, 62).

Public and private interests, the property boom and Bankgesellschaft Berlin

Potsdamer Platz is probably the starkest example of the aspiration to become a global city; the material proof of the ambition to re-build and, in the process, re-define Berlin as a global city. This massive construction project, employed world famous architects to design high-rise buildings for global brands like Sony to work in, and provide bars, restaurants and cinemas for visitors (Cochrane and Passmore 2001, 349). The Potsdamer Platz construction project both contributed to, and was symbolic of, the economic hype in Berlin. It was, however, only one part of the largest construction site in Europe at the time.

From 1990 Berlin had been subject to the interests of property developers. Large international and national real estate firms expected continued growth in demand for office and retail space in Berlin. Following the logic, 'if you build them, they will come', Berlin's real estate boom resulted in vast empty office spaces. The notion that Berlin would somehow move quickly from the Cold War to the centre of the global economy appears to have acquired an almost commonsensical air. One of the defining features of discussions about Berlin was the hope, indeed, hype that persisted through the decade. It was fuelled by speculation, as those scenting a development project, contributed to the boom. There were, for example, forty applications submitted to the Brandenburg regional authorities (the region surrounding Berlin) for golf courses and twenty-seven applications for Disneyland-style parks in the 3 years immediately following reunification (Ellger 1992, 45).

However, as early as 1992, some concerns were being raised about Berlin's capacity to develop into a prosperous capital and global city. Concerns did not just rest on the socio-economic capacities of the city but also the capabilities of the politicians, viewed by some as "too inward looking to meet its global city status" (Brügel 1993, 156). Such thinking was apparent in the Federal government's measures to curtail the Berlin government's responsibilities (Brügel 1993, 156). The decision-making of Berlin's politicians and its advisors should be seen as a major factor in the emergence of Berlin's problems in the 1990s. Krätke (2004a, 2004b) has convincingly argued that, regardless of the challenges Berlin may have faced, the policies of the CDU-SPD coalition government contributed to the severity of the fiscal crisis which emerged in the mid-1990s. The clearest example of this is probably the continued financial and political support for property speculation beyond the end of the mini-boom in the mid-1990s. The main means through which the government did this was the Bankgesellschaft Berlin and a strategy of urban marketing. The former was particularly emblematic of politics in 1990s Berlin. It was a public-private partnership, it captured the crucial role played by consultants in developing policies and it ended in failure and corruption.

The Bankgesellschaft Berlin was a public-private financial corporation, established in 1994. A key aim of the corporation was to encourage investment and provide loans in the property market of Berlin and Eastern Germany. The complicated legal framework, known as the 'Holding Model' bound public and private interests together in the Bankgesellschaft and was to become the basis for the BWB partial privatisation model.[22] Consultants designed the model for this partnership. They can in this sense be seen as the designers of new governance structures. In the terms of governmentality literature, it was the consultants who provided the practical means – the techniques of government – through which the ambitions of the government could be realised.

Public-private partnerships and consultants were also apparent in the marketing and development of Berlin *e.g.* Wirtschaftsförderung Berlin (Business Location Marketing), Berlin Tourismus Marketing and Partner für Berlin (Capital City Marketing) (Häussermann and Colomb 2003, 203).[23] These organisations also revealed an apparent merging of interests between public and private in Berlin during this time. Partner für Berlin, for example, was established in 1994 with 50% funding from the government and 50% from major German companies and local businesses. Though the city was not officially a shareholder, it commissioned Partner für Berlin to design and implement marketing strategies for Berlin. Furthermore, the Chief Executive was a former Senator for Urban Development in Berlin still well connected to the Berlin government as well local businesses (Häussermann and Colomb 2003, 203).

---

[22] The study will return to this model in the detailed discussion of the BWB privatisation process in Chapter 6.

[23] Häussermann and Colomb's (2003) translations of the company names.

These consultancy firms were crucial to the re-imagining of the city, of making the 'New Berlin' attractive to investors and visitors. Working with public and private money, they became integral to the strategy of re-making Berlin. Another clear example of this ambition was the – failed – application to host the Olympics in 2000. This was branded as an important step towards becoming a global city, in assisting the socio-economic restructuring of the city and the general economic development of Eastern Germany (Brügel 1993, 158). Such hype persisted despite the everyday socio-economic realities. Indeed, contradiction between rhetoric and reality became a feature of political discourse and was still strong at the end of the 1990s: "Despite the ongoing economic recession, high unemployment and fiscal crisis, Berlin boosters talk about the metropolis re-emerging as a high-tech, information-oriented global city that opens up eastern markets" (Campbell 1999, 174).

With regard to the property market and urban development, the outcome of this boosterism was some of the largest construction projects in the world. Bankgesellschaft played an active role in speculative investment not just in Berlin but across East Germany despite the fact that in 1993, the boom in East Germany and Berlin had turned into a crisis with an over-surplus of space and declining rents. The boom was, at least, to some extent "politically driven" (Krätke 2004b, 522) in that it was instigated by the Federal government's establishment of a tax write-off incentive for investment in property in Eastern Germany. It did, however, merge "perfectly with the urban government's belief that office building sites are a sign of economic progress and a promising future" (Krätke 2004b, 523).

Alongside this blind optimism and perhaps inevitably within the 'old boys' network of 1990s Berlin politics, corruption was apparent. The Bankgesellschaft became a key player in the speculative real estate bonds in the Berlin and wider East German market. In the process they created special 'VIP' bonds which were offered exclusively to "well-known representatives of Berlin's political class and urban government" (Krätke 2004a, 62; for more detail Rose 2004). This was revealed in 2001 in the so-called 'Banking Scandal' when it transpired that leading politicians from the then largest party, the CDU, had profited from these funds. This led to the collapse of the CDU-led coalition, the launching of criminal proceedings against prominent Berlin politicians and the shattering of public confidence in the political establishment. Ultimately, the Berlin government was forced to intervene and guarantee the Bankgesellschaft's debts and – controversially – payments to the bond-holders. In the 1990s Berlin acquired the "dubious reputation of being administered by a corrupt elite" (Eckardt 2004, 6).

Problems had become apparent as the property crisis (of low demand, empty spaces and declining rents) deepened through the 1990s. Of course, the Bankgesellschaft had contributed to the massive surplus of office and retail space with its financial support for property developers. Berlin had become symptomatic of "urban capitalism's weak underbelly tendency of credit-financed over-building" (Ward 2004, 246). The speculative bubble, with the promise of tax write-offs and the generous financial support of Bankgesellschaft encouraged huge investment in property development projects across Berlin. In combination with the drop in population, rising unemployment and welfare claimants, these tax breaks contributed to the decline in the city's tax revenues (DIW, Deutsches Institut für Wirtschaftsforschung 1997 in Krätke 2004b, 62).

The global city discourse: translating neo-liberalism in Berlin

Bankgesellschaft Berlin and the urban marketing of Berlin are emblematic of policy-making in the 1990s. They illuminate some of the effects of the global city policy discourse as it was translated into policy practice by the Berlin government. They reveal the translation of neo-liberalism and new forms of governance in the context of Berlin: the interplay between broader trends and local politics. The outcomes of policies and the actual shape governance structures took were contingent on actors in Berlin but they were reflective of broader trends in policy-making. The utilisation of public-private partnerships in urban governance, though long apparent, grew considerably in Europe in the 1990s and is representative of the shift to governance (Stoker 1998b, 34). In urban planning, this trend has led to criticism that decision-making processes are increasingly privatised, with consultants in particular seen as powerful in shaping urban governance (McCann 2001; Martin *et al* 2003). In developing the model for the Bankgesellschaft and playing prominent roles in the urban marketing public-private partnerships, consultants were at the heart of decision-making processes. They were integral to the new, informal and ad hoc 'governance' arrangements through which public and private actors made policy.

The government may have implemented policies which were common to many cities in the period, but they did so in their own particular fashion. Krätke states that "Berlin's economic policy in the 1990s is an outstanding example of 'worst practice' urban governance, since it has led to a financial crisis with truly catastrophic effects" (Krätke 2004b, 525). The 'Banking Scandal' can, in some ways, be seen as the continuation of the political practices of an elite shaped by the peculiarities of West Berlin's subsidy-rich, Cold War isolation (Rose 2004). In the 1970s and 1980s there were a series of remarkably similar, interlinked scandals (*Steglitzer Kreisel*, *Dietrich Garski* and the *Antes* scandal) in which politicians, construction companies and financial institutions had shared subsidies between themselves (Chacon 2005, 107). As in the Banking Scandal of 2001, these cases of corruption led to the resignation of the city's Mayor (twice), provided profits for the companies involved and left the State of Berlin to pay for the losses incurred

(Chacon 2005, 39).[24] In a sense, then, although the context of policy-making had changed (from subsidised Cold War isolation to neo-liberal globalisation) certain characteristics of governing in Berlin remained apparent. The fact that this decade also witnessed a large swathe of privatisations across a range of sectors (Electricity, Gas, Housing as well as Water) means that the two subjects were entwined with each other (see Passadakis 2006). Privatisation may already have been associated with corruption in other parts of the world, but in Berlin there were very local reasons for assuming that it might lead to corruption.

A politics of inevitability

Global city policy-making thus had both material and discursive effects. It was a way of thinking about policy-making – seen in the production and re-production of an increasingly neo-liberal way of constructing policy problems and solutions. It was also a way of practicing politics – seen in the production and re-production of public-private partnerships in the making of policy and as the objective of policy. In this sense the material and discursive effects of the global city discourse should be understood as a process of translation: the "simultaneous production of new knowledge and a network of relations" (Callon 1986, 203).

The policy discourse can also be seen to have had material effects in the sense that it generated policies in 1990s Berlin; polices which were not always successful. Still, as highlighted above, global city policy-making continued, even increased, through the 1990s. A cyclical dynamic of cause and effect is observable: one in which the failure of one neo-liberal response to the challenges of reunification and globalisation, leads not to a re-consideration of that policy, but to the translation of another neo-liberal policy. A politics of inevitability emerged. Crucial to understanding this dynamic was the potency of the storyline of inevitable and necessary re-structuring of the economy. This served to justify the tensions between the symbol of the global city and the socio-economic realities of 1990s Berlin. As economic and fiscal problems mounted, the rationality of the policies – or those making them- was never wholly re-problematised. Rather, following the global city storyline, these effects were seen as necessary hardships. What was required were new policies to make Berlin a global city within the changing context. Thus the government's policies, as framed by the global city discourse, contributed to the problems of Berlin. The challenges of reunification and re-integration may have been great, but the CDU-SPD government of the 1990s bears much responsibility for the dire state of affairs Berlin found itself in at the end of the decade.

---

[24] See Beveridge and Hüesker 2007 for a more detailed overview of corruption and Berlin politics.

The next section of the chapter examines how a politics of inevitability emerged in the WSS sector. New neo-liberal forms of governance also emerged in the WSS sector: the utilisation of public-private partnerships; the 'need' for private sector investment and private sector experts to realise the global city aspiration. It is argued that the privatisation of BWB had little to do with debates over water management, especially of technical and environmental aspects. Instead, the privatisation reveals the extent to which water policy-making became wrapped up in the broader political objectives of the government: how decision-making regarding BWB became subject to the terms of the global city policy discourse and the material effects it produced. Ultimately, it shows how the structures and logics of water policy-making and management became re-represented and re-assembled through the translations of the global city policy discourse.

Translating the global city policy discourse in the water sector

Soon after reunification the magistrate of East Berlin decided to transfer the responsibility for managing the East Berlin water and wastewater operator VEB Wasserversorgung und Abwasserbehandlung Berlin (WAB) to BWB, though the two companies only fully and formally merged in 1992 as a municipal utility (Eigenbetrieb) (Lanz & Eitner, 2005a, 10). The reunified water company faced a number of challenges: there was a need for investment in infrastructure, largely in the East of the city (Lanz & Eitner, 2005a, 10); a surplus of workers (Interview 8); and the perception that the system did not have the capacity to cope with the anticipated rise in levels of consumption as Berlin 'boomed'. As a result of this latter perception, local water and environmental consultants had some success in promoting the expansion of infrastructure in the city (Interview 10). However, all in all, in the early and mid-1990s neither a surplus of employees nor a need for investment in infrastructure was seen as a major problem. Indeed, BWB was not discussed as a burden on the city's finances during this time and privatisation was not presented as necessary to raise funds to ease the city's debts (Interview 1).

Still, privatisation of BWB was discussed as early as 1992 (Lanz & Eitner, 2005a, 10). Proposals came from within the CDU leadership and rather than focusing on a real 'need' for privatisation, as occurred in the UK in the 1980s, the emphasis was that privatisation was 'more efficient', a 'superior' way of managing utilities (Interview 1). Ultimately, however, the case for privatisation made little headway in the early 1990s. With expressions of public opposition to the policy, plans were dropped in 1993 (Lanz & Eitner, 2005b). It was not until after the success of the CDU (and poor performance by the SPD) in the election of 1995 that privatisation of BWB, (and of the publicly owned Gas and Electricity companies) became an official policy objective in the CDU-SPD Coalition agreement (Koalitionsvertrag). It did not, however, become an official policy proposal until 1997. By this time the fiscal situation in Berlin had worsened, in part as a result of the end of the mini-boom and the emerging property market crisis. Furthermore, BWB had been transformed from a profit-making into a loss-making company between 1995-1997.

A 'fresh wind' through Berlin

If privatising BWB was pushed to the margins of the policy-making debate in the early and mid-1990s, water governance was more generally becoming framed by the global city policy discourse. 1994 marked a key point in the process: the decision to legally enable BWB to act as a profit-making company, through converting it into a public law corporation. Along with the state-owned Electricity and Gas companies, BEWAG and GASAG, BWB was converted into an Anstalt des öffentlichen Rechts, (AöR).[25] This is a much looser legal status, one which allows public companies greater freedom to act independently from government (Lanz & Eitner, 2005a, 24) and to pursue commercial objectives.[26] Freeing up these public companies was supposed to bring a "fresh wind" (Rose 1998) through the city. The main aim was to encourage entrepreneurship by allowing these companies to act commercially outside of Berlin. The official justifications for this change were that BWB needed extra funds to carry out infrastructural improvements in Eastern Berlin and to reduce the workforce employed by the municipality (Lanz & Eitner, 2005a, 10). However, a diverse range of interviewees stress that the decision was driven less by necessity and more by a desire to make BWB a player in the international water market (Interviews: 2, 3, 4, 8). This was a common strategy in the water sector with both privately and publicly owned companies, such as those in Turin and Seville, expanding their operations geographically and becoming embedded in international competition (Swyngedouw 2003, 8).

BWB was actually a profitable company at this stage. In 1995, the total revenue of BWB was $1 Billion (DM1.75 Billion) and the profit for that year was around $30 Million (DM 51 Million), according to the Executive Director (Frankfurter Allgemeine Zeitung 1997a). In other words, in 1995, when the reform came into effect, BWB was not seen as badly managed or a company in trouble. In fact, according to a member of the BWB Management Board at the time of privatisation it had a good reputation (Interview 3). This step was therefore more of

---

[25] An Anstalt öffentlichen Rechts (AöR) is a public law institution concerned with the implementation of public functions and responsibilities. These duties are assigned to it by law. It has been utilised by Länder/ federal states and the Kommunen/ municipalities.

[26] This is achieved through removing the Örtlichkeitsprinzip which binds public companies to activities only in their local domain.

a re-definition of the company's purpose. The hope was that BWB, with its apparent expertise, would be able to exploit the perceived opportunities emerging in the international water market, particularly in developing countries and Central and Eastern European countries. The aim then, as upon privatisation in 1999, was to make BWB an international centre of water management expertise (Berliner Morgenpost 1998).

This marked the beginning of a neo-liberal re-making of BWB. It was no longer seen as simply being a public company which produced and supplied water to domestic and commercial users in Berlin and treated and disposed their wastewater. BWB was to become a commercial player, a profit making organisation, acquiring prestige for itself and Berlin. With this step, water policy-making was re-constituted in the terms of the global policy discourse. BWB was embedded within the storyline of 'necessary and inevitable re-structuring' as translated in the water policy and management sector. The commercialisation of BWB was a response to changes in the global economy. A policy vocabulary centred on 'competitiveness' emerged along with an aspiration to global success. Just as Berlin was to be re-made as a global city, BWB was to be re-made as a global player.

Upon making the decision, the government appointed a new Executive Director, Bertram Wieczorek (who remained until privatisation in 1999). He was a prominent CDU member, having been Secretary of State in the Federal Environment Ministry under Helmut Kohl. Reforms may have been aimed at making BWB – officially at least – more independent, but the management of the company remained distinctly political. In fact, a workers representative at the time argued that Wieczorek was under some pressure to help realise Kohl's vision of a revitalised Eastern Germany (of *blühende Landschaften* – "blooming landscapes") and this accounted for many of the mistakes made during the mid- to late-1990s (Interview 10).

The strategy for making BWB a global player rested in large part on the ultimately flawed notion that Berlin was a 'gateway' city to emerging markets in Eastern Germany and Central and Eastern Europe. Thus though Wieczorek may have "had no idea how to run a water company" (Interview 8), and many of BWB's investments would prove to be loss-making failures, the strategy itself and the means through which this was supposed to occur, reflected the broader assumptions and strategies underpinning policy-making during the period. The rhetoric was also similar. Just as Berlin was hyped as an emerging 'high-tech' global city, the conversion of BWB into an AöR and its high profile investments in the mid-1990s were seen as having made it – at least by 1997 – not just "profitable" but "above all innovative" (Füller 1997). One investment, the purchase of *Schwarze Pumpe* (SVZ), was at this stage hailed as the acquisition of the most advanced wastewater treatment plants and BWB was presented as the "pearl" of the municipally-owned companies (Füller 1997).[27]

Commercial problems: the emergence of the 'need' for private sector expertise

The pace of expansion could clearly be seen by mid-1997 when privatisation publicly emerged as a policy option. BWB had become the market leader in terms of customers in the Brandenburg water sector (the region surrounding Berlin) and had subsidiary companies in Poland, the Czech Republic and Russia (Fischer 1997). If the objectives of water policy-making were re-framed through the global city policy discourse, the actual outcomes of decision-making in this period also bore similarities to those of Berlin's more general re-development. There was a clear and apparent contradiction between the claims made about the commercialised company (how it was represented) and the performance of the company.

The purchase of *Schwarze Pumpe* (SVZ) near the Polish border in Eastern Brandenburg in June 1995 was emblematic of this process. Initially praised, this investment became the most high-profile failure amongst many and became a key problem for the Berlin government in the privatisation negotiations. In December 1997, Rose and Sontheimer (1997) reported in Der Spiegel magazine that (SVZ) *Schwarze Pumpe* had made losses of $89 Million (DM 155 Million) between 1995-1997. Although they accepted that the costs of reunification had also played a role, Rose and Sontheimer (1997) saw a direct correlation between such failed investments and the fact that water prices in Berlin were the highest in Germany. BWB management was forced to deny these claims, stating that investments were funded by Bank loans alone and that price increases were the result of the costs of reunification especially the upgrading of infrastructure in East Berlin (Wiengten 1997).

Costs were incurred through upgrading but it is hard to argue that they alone accounted for the need for higher prices. BWB was a profitable company, at least prior to commercial expansion. As stated, in 1995 the company made a profit of around $30 Million (DM 51 Million) (Frankfurter Allgemeine Zeitung 1997). The perception was, however, that it would make much more of a profit if it expanded into water markets. The profit-making capacity of BWB was also becoming increasingly important to the government. The city's fiscal problems increased between 1995-1997, with debts rising from about $27 Billion (DM 46 Billion) in 1995 to around $33 Billion (DM 56.46 Billion) in 1997 (Marschall *et al.* 1997). Given this context it is likely that the government pressured BWB to increase profits. Perhaps Wieczorek's remarks in September 1997 confirm this. He stated that BWB

[27] Sekundärrohstoff-Verwertungszentrum Schwarze Pumpe GmbH.

must further reform itself, if price increases were to be avoided; hence, he argued for expansion plans to countries such as Namibia, Chile, China and Turkey (Der Tagesspiegel 1997c).

This re-making of BWB as a commercial company was part of a wider trend in which publicly owned water companies were forced to operate in a "ring-fenced manner, as autonomous, self-financing organisations that operate according to market logic" (Swyngedouw 2009b, 39). More generally, the problematisation of public companies' status and functions also resonates with the long-running saga over the privatisation of the German Railways Company, Deutsche Bahn. Like BWB, Deutsche Bahn's reputation with its original consumer base (domestic users of rails services) has declined as it has increasingly sought to internationalise. As with BWB, one of the main arguments put forward for a partial privatisation was that private sector expertise and investment were required to ensure that the company became a success in the soon to be liberalised European rail sector (Gow 2009).

If the objective of re-making BWB as a global player and the strategy of investing in Central and Eastern Europe bore the marks of the broader global city discourse, the policy practices of BWB during this period also share similarities with the broader practices of the policy discourse. First, the prominent role played by consultants in decision-making. A Union representative at the time stated in interview that consultants were instrumental to the management of BWB between 1994-1999 (Interview 17). A senior Left Party politician in Berlin was even more specific, stating that the consultants' key role was in advising on investments (Interview 7). As well as acting as a commercial player in water markets outside of Berlin, BWB invested in other sectors, diversifying even within Berlin. According to this Left Party politician, consultants were behind BWB's investment in the telecommunications company Berlikomm, established in 1997 (Interview 7). One aim of Berlikomm was to market BWB's pipe network as the basis for the development of a *Hochmodern* ('state-of-the-art') data network in Berlin (Frankfurter Allgemeine Zeitung 1997). This attempt to re-make BWB as a commercial player and Berlin as a high-tech city would almost predictably become another loss-making venture.

If the use of consultants was entirely in-keeping with the Berlin government's reliance on the private sector for the bigger project of re-making Berlin, then so was the hint of corruption. Once again the *Schwarze Pumpe* (SVZ) case seems to capture this quality. Rose and Sontheimer (1997) and Rose (2004) thought that BWB's investment in *Schwarze Pumpe* (SVZ) was linked to the SPD Federal Politician Ditmar Staffelt. Staffelt had strongly supported the appointment of the CDU politician Wieczorek and had interests in *Schwarze Pumpe* (SVZ).

Proposing the privatisation of BWB, 1997- 1998: a politics of inevitability?

On the 24th June 1997, *der Koalitionsausschuss* (the Coalition Committee) of the CDU and SPD, formally declared that privatisation of BWB should be discussed. This was part of Fugmann-Heesing's 'asset mobilisation' programme announced in March 1997, in which the major state companies were to be considered for privatisation in order to raise revenue. By the end of the summer of 1997, members of both the SPD and CDU leaderships had openly called for privatisation of BWB. If there was, as yet, little agreement between the CDU-SPD governing coalition on what form privatisation should take, there was apparent agreement on the necessity of privatisation. Superficially at least, the ready-made reasoning for this was clear: the government needed money from the sale. Such a straightforward explanation is, however, problematic.

Later in 1997, BEWAG, the electricity company majority share owned by Land Berlin was privatised and in 1998 GASAG, the publicly owned gas company was also privatised. Though hardly popular, privatisation in these sectors (where the private sector already played a prominent role) would not have been seen as such a risky policy as the BWB privatisation. Privatisation in the water sector is far more challenging and controversial, as a SPD politician in the government admitted (Interview 12). Furthermore, in the German context, privatisation of water services was uncommon. At this time, and to a lesser extent still today, there was a good deal of consensus around the notions of the social market, as expressed in the concept of the *Kommunale Daseinsvorsorge* and illustrated in municipal ownership and operation of most water utilities. Accordingly, it is wrong to assume that the formal proposal to privatise BWB would simply have been seen as a rational, logical option, even within the local context of a fiscal crisis and the global context of the rise of privatisation. Privatisation of BWB did therefore have to be produced as a rational and inevitable policy option.

Freeman (2007, 3) states that the policy-making process is given an appearance of simplicity by the formal procedures of democratic systems in which discussions are announced, parliamentary debates held and decisions officially made. Policy appears – formally at least – describable in terms of stages or cycles. The reality is, of course, far more complex. Decisions "are prefaced by lengthy periods of establishing what does and what does not need to be decided upon, and what the parameters of any decision might be" (Freeman 2007: 3). This chapter argues that the – discursive and material – parameters of the decision to privatise BWB were set through translations of the global city policy discourse. They structured the debate through the "mobilisation of bias" (Hajer 2002, 62); serving to shape the options seen as available to actors and prompting the definition of 'problems' and appropriate 'solutions'.

Here the notion of translation is useful. Seen from a translation perspective, the proposed privatisation of BWB marks the beginning of a process of re-representing and re-making BWB. It problematises BWB's status: it is failing to become a 'global player'. It also brings BWB into relation with the fiscal crisis: Berlin needs money so BWB should be privatised. It is about both the creation of new knowledge and new relations between actors. The problematisation of BWB introduces new knowledge claims and threatens the status of some actors, while offering opportunities to others. Viewing it from the translation perspective, it is the attempt to "substitute one set of relationships or associations with another" (Freeman 2007, 5). It is the attempt to impose a new order through the (re-) representation and (re-) association of institutions, individuals and objects (Freeman 2007, 5). As Callon states a "problematisation possesses certain dynamic properties: it indicates the movements and detours that must be accepted as well as the alliances that must be forged" (Callon 1986, 206).

The problematisation was informed by a neo-liberal rationality – it was informed by particular assumptions; it had a normative quality. In governmentality terms, "the practice of government involves the production of particular 'truths'" (Larner and Walters 2004, 2). The proposed privatisation of BWB makes claims about the present and about the future – it aims to establish truths about Berlin and BWB which together make the 'no alternative' case. Relating Berlin to BWB (through reference to the terms of the global city policy discourse), the problematisation seeks to set the terms of the debate – to establish a sense of there being 'no alternative' to privatisation.

Thus the government's proposal to privatise BWB in 1997 should not be seen as being based on an objective assessment of the situation. "Problem definition is a matter of representation" (Stone quoted in Wilson 2000, 254); it is the result of a particular way of interpreting the relationships between things. In other words, the policy discourse acts as a means of interpreting the growing fiscal crisis and the emerging problems with BWB's investments; themselves, in part, material effects of global city policy-making. Thus the policy discourse and the policy-making it informed are fundamental to the material and discursive conditions in which the BWB privatisation emerges. It is a neo-liberal reading of the situation – a particular means of understanding the role and 'value' of BWB.

As potent as the global city storyline was, there would still, of course, be opposition to privatisation. As stated in Chapter 4, a policy discourse does not magically make actors do things. The Berlin government and their advisers aimed to create a sense of inevitability about the privatisation and the financial context may have added to this sense. However, the privatisation was not *per se* inevitable – it had to be constructed as such. It is thus ultimately about actors and local, context-specific processes of coercion, bargaining and confrontation.

It is at this point that the emphasis in the analysis shifts: from tracing some of the key translations of the global city policy discourse to following actors translating discourse into practice. To move the focus of the analysis onto the actors is to appreciate the importance of institutions and the formal processes of policy-making; to remember that both the 'politics of organisation' and the 'organisation of politics' are crucial (Gottweiss 2003, 254). It is also to understand that policy can be seen as the creation of a "collective script" (Freeman 2007). This script may have been similar to many others being written around the world at this time, but the actual acceptance of roles and the coordination of performances in the privatisation required the support and agreement of actors in Berlin.

In other words, the problematisation of BWB may have been set-up by the translations of the global city policy discourse but the privatisation still had to be achieved. In stressing the potency of this neo-liberal discourse, the chapter does not suggest that actors had no room for manoeuvre. Here the tension between discourse and agency is evident in the study. On the one hand, the policy discourse almost seems to explain political agency: the apparent 'inevitability' of the privatisation. On the other, political agency produces the discourse: the inevitability needs to be produced and the privatisation realised.

This tension can be further illuminated with a discussion of the Finance Senator, Fugmann-Heesing. When the privatisation of utilities emerged more strongly in the mid-1990s, there had been a clear shift within the SPD (at least in the leadership of the Parliamentary Party). The SPD had suffered defeat to the CDU in the Berlin election of 1995 as Berlin's economic problems and unemployment rose in the 1990s. In response, the right wing of the SPD grew more powerful and argued that the party must consolidate its position in the centre to win back votes from the CDU (Interview 7). This mirrored changes within the SPD at the national level, as Gerhard Schröder with his notion of the *Die Neue Mitte* ("The New Centre") and Third Way politics emerged after yet another defeat to Helmut Kohl in the national elections of 1994. Within this context, the SPD in Berlin appointed Fugmann-Heesing (January 1996). She was, according to almost every interviewee, a crucial driving force for privatisation. Interestingly, being parachuted in from the state of Hessen, she was something of an 'outsider', and certainly not a member of the political club in Berlin. She was brought to Berlin by Klaus Böger, the SPD Group Leader (and by this stage a supporter of privatisation) to push through what he saw as the hard but necessary reforms of governmental policy (Interview 7).

As well as being a key actor in the privatisation and Berlin politics in general, Fugmann-Heesing's arrival can also be seen as emblematic of broader shifts in the period. Ultimately, the study does not make a realist attempt to explain her motives, or establish a cause and effect sequence. Instead, the contingency of the process is stressed: the sense that the global city policy-making discourse was made to exist only through the exchanges between actors. Furthermore, the agency of figures such as Fugmann-Heesing is seen as being contingent upon these negotiations too. Despite her reputation as the 'Iron Lady' (Interviews: 9, 14), she did not simply exert her will or exercise her official powers and impose the privatisation. In fact, the eventual partial privatisation can be seen as a complex compromise. It was, in part, a reflection of the Berlin way of doing politics; of, in particular, powerful Unions and coalition politics between the SPD-CDU.

The policy process 1997- July 1998

The government's official declaration (24th June 1997) that BWB should be privatised followed a period in which the privatisation of BWB had been increasingly discussed. The civil servant at Finance stated in an interview that it emerged in 1996 as a serious issue for the leadership. By 1997, the fiscal crisis in Berlin had increased and a survey of newspaper sources from spring through to the end of the year reveals heightened debate over BWB's future. In May 1997 Wieczorek claimed that he had information that the Berlin government would sell BWB (Berliner Zeitung 1997). In June, the Berlin government confirmed that there had been contact with Suez regarding a "quick sale", though Suez denied rumours that they were prepared to offer around $1.78 Billion (DM3 Billion) (Berliner Zeitung 1997). According to other press reports in this period, many international water companies and German energy companies had already signalled their interest in BWB (Der Tagesspiegel 1997a; Handelsblatt 1997).

There was open discussion in the media regarding what form the privatisation should take amongst those who supported it and what could be done to avoid it amongst those against it. The Chairmen of BWB, Wieczorek, proposed a flotation on the Stock Market as an alternative to privatisation (Frankfurter Allgemeine Zeitung 1997b). The former communist party, PDS, even proposed putting it on the Stock Market with a majority share remaining public as an alternative both to privatisation and the removal of $1.2 Billion (DM2 Billion) from BWB's capital (Berliner Zeitung 1997c). The source of this latter proposal was the accounting firm KPMG who were advising the Berlin government at the time (Berliner Morgenpost 1997).

Such was the speculation about BWB, coming in the context of the ongoing privatisation of BEWAG (1997) and the preparation of GASAG for privatisation (in 1998), that a representative of the Union (ÖTV) spoke of a *Schlussverkaufsmentalität* (a "closing-down sale mentality") in the city (B. Z 1997).[28] Adding substance to such fears was Fugmann-Heesing's decision to sell public

[28] The Union representing workers of public companies such as BWB.

buildings and property despite opposition from within her own party. By 1997 Fugmann-Heesing, usually with Böger's support, was prepared to propose unpopular policies such as privatisation to reduce the interest being paid on the deficits, if not the deficits themselves. These were set to top $2.7 Billion (DM 4.68 Billion) by 2001 (Berliner Zeitung 1997b).

In his ready-made version, the civil servant at Finance made a clear distinction between the real reasons for privatising (Berlin's debts) and the – neo-liberal – rhetoric used in public (the need for private sector expertise, etc.). Others supported this distinction, as if to emphasise that the decision was objective, born solely of necessity. For example, one of the consultants who worked for the Finance Senate argued that there was a "strong need to privatise" and that it was not an "ideological" venture (Interview 9). A reflection on the discursive context of the privatisation debate, suggests that drawing any distinction between the 'rhetoric' of private sector expertise and the harsh realities of the fiscal crisis is artificial. The BWB privatisation may ultimately have been driven more by a need to raise finances than a desire for the expertise of the private sector. However, these two arguments were bound-up with each other in the problematisation of BWB, informed by a neo-liberal rationality in which private sector expertise and fiscal prudence were part of a more fundamental promotion of the market and the reduction of the state. Furthermore, moves to privatisation were increasingly apparent and accepted around the world in this period. The 'no alternative' argument gained much of its potency from the prevalence of privatisation around the world.

Emphasising the importance of the global ascendancy of the neo-liberal discourse does not mean that its effects should be assumed. Here a sense of the interplay between discourse and agency is required. To repeat the quotation from Peck and Tickell (2002, 53), neo-liberalism is not an "actorless forcefield of extralocal pressures and disciplines". Rather it was produced through local political strategies and actions. The comments of the Berlin-based journalist may be instructive here: rhetoric about the need for private sector expertise was, in fact, used to distract from genuine debate about the true causes of Berlin's parlous finances and its desperate need for capital (Interview 4).

This is a plausible explanation for the emphasis put on the superiority of the private sector. By setting the terms of the debate in such a way, it was easier for the government to avoid a discussion of its own role in the financial woes of Berlin and the poor performance of BWB. The definition of problems in policy-making is often strategic because a government can "deliberately and consciously design portrayals so as to promote their favoured course of action" (Stone in Wilson 2000, 254). Political actors can and do shape and manipulate discourses for their own purposes.

The journalist went on to state that BWB's disastrous commercial phase, as well as accruing debts, also fuelled arguments about the need to bring in private sector management to make it commercially successful (Interview 4). As early as the summer of 1997 it was alleged that a secret report produced for the government by *Bossard Consultants* (who had also worked on the failed Olympic Games bid) claimed that water prices in Berlin were too high and were the direct result of the BWB Management Board's poor decision-making in water markets (Riecker 1997). Absent from this report was the role of those consultants advising the BWB Management Board on these investments. The focus was placed directly on the public management. The role of private sector consultants in BWB's problems was not problematised. Instead, the publicly owned, in part politically managed, BWB was responsible for the failures.

While accepting that the arguments for privatisation were often knowingly strategic, the responses of other actors were suggestive of the subtle ways in which neo-liberalism had shaped the terms of the debate. There is a sense that certain neo-liberal assumptions about BWB had become taken for granted, even 'black boxed', through the prior translations of the global city policy discourse. For instance, though not all participants in the policy debate accepted the *Bossard Consultants'* claim about the need to bring in private sector "know-how" (Riecker 1997), the objective of becoming a 'global player' was never seriously problematised. As is clear in the Table 4.1 below, even opponents of privatisation, such as the Unions, accepted that BWB must continue its commercial activities – that it must raise money for Berlin. Thus all alternatives to privatisation implicitly accepted that BWB should continue to act as a 'global player' despite the fact that the pursuit of this objective had transformed BWB from a profitable to a loss making company.

It could simply be argued that this was the result of the fiscal crisis in Berlin – that actors accepted that BWB must be used to raise income. However, such an argument overlooks the way in which the global city policy discourse relates BWB to the fiscal crisis, the way in which it sets the terms of the privatisation debate. Hajer and Fischer's (1999) statement regarding the potency of the ecological modernisation discourse could be seen – by way of analogy – as applicable to the global city policy discourse in Berlin: "In order to be heard, one needs to comply with the terms of this pre-given discourse" (4).

The privatisation proposal should not be seen as the outcome of a rational, objective assessment of the situation by the government. Rather, the terms of the global city policy discourse mobilised bias. 'Facts' became mixed with values. For example: the 'fact' that BWB was not succeeding as a global player – by now increasingly apparent – leads to the solution that private sector expertise is necessary to make it succeed. The failure of commercialisation does not, in this problematisation of BWB, lead to a re-assessment of the objective itself – why should BWB become a global player? Having been translated, it remained black-boxed: the objective itself and the strategy through which it should be achieved are not seriously contested in the light of BWB and Berlin's growing problems. The commercialisation of BWB can be seen as a key translation of the global city policy discourse into the water sector, the reform of the company's objectives and management in the context of neo-liberal globalisation. In some ways, it can also be seen as the first step to privatisation, narrowing down future options, in the sense that as BWB was acting in some arenas as a private company with private sector logics and objectives.

Table 4.1. Policy proposals in the privatisation debate.

Model	Key Supporters	Key Opponents
"Holding model": Combination of AöR (core business officially remaining public law company under city control) and holding with silent partnerships; private investors take over operations, strategic decisions cannot be taken without their consent; investors hold shares of assets as well.	Consultant F.-J. Pröpper, CDU group (majority), SPD Parliamentary group, SPD Berlin (majority, internal debates).	ÖTV, BWB staff council, BWB staff, Greens, PDS.
"Conversion model": BWB to be converted into joint-stock company with BWB shares sold to non-water investors (banks, insurance companies, pension funds etc.).	Branoner, parts of CDU and SPD Parliamentary party.	ÖTV, BWB staff council, BWB staff, Greens and PDS.
"Integration model": Merger of municipal companies (in this case, water and gas suppliers) to form a public law multi utility company; at a later stage, this was combined with a concession model for the withdrawal of capital.	ÖTV, BWB staff council, BWB staff, BWB management (minority), SPD Parliamentary group (majority, until late swing).	CDU group (majority), SPD group (minority), Fugmann-Heesing and Diepgen.
"Concession model": Model based on long-term lease of BWB (20-25 years), concession fees to be financed by credit the costs of which would be born by the consumer.	Suez, Fugmann-Heesing (for a time).	CDU group, SPD group (majority), Branoner, ÖTV, BWB staff council, BWB staff. PDS and Greens.

Source: adapted from Lanz and Eitner 2005a.

Is it, however, possible to imagine that had BWB been successful these new arrangements could have been maintained? Could it have remained a publicly owned and operated water and wastewater company within the boundaries of Berlin and a commercially active company outside of Berlin? According to contrasting sources, yes. Both the member of the BWB Management Board and a Left Party politician, argued it was not the actual decision to commercialise, but the disastrous performance of BWB after the event that made privatisation virtually inevitable (Interviews: 2, 3). Both believed that the decision to privatise was primarily driven by the continued desire to make BWB competitive in international markets. Thus, in their view, had the venture been successful, privatisation may never have come on the agenda, even in the context of the growing fiscal crisis. This is debatable, of course, given the global trends in WSS management, but not impossible when the general lack of water privatisation in Germany is considered.

More fundamentally, in the view of these two contrasting figures, the privatisation was not merely about reducing Berlin's debts. The privatisation did not arise merely from an objective assessment of Berlin's fiscal situation. It was, instead, indicative of how the assumptions and strategies of a neo-liberal political rationality defined policy objectives. It can be seen as part of a broader project of re-making Berlin. This may sound a little far-fetched, given the fiscal situation, but a leading figure from the SPD, working at the Finance Senate, almost said as much.

According to this politician, the privatisation of BWB emerged on the policy-making agenda as part of their broader reforms aimed at tackling Berlin's budgetary problems: to reduce government expenditure and to reduce debts through bringing in capital (Interview 12). The financial arguments for privatisation, in particular that the surplus of public workers in Berlin was acting as a drain on the government's expenses, were supported by other interviewees (Interviews: 8, 11, 14). However, the SPD politician stated that there was a broader, deeper objective to their reforms. They wanted to "change the philosophy of Berlin": to rid Berlin of what they saw as its dependency culture born of Cold War subsidies in the West and socialism in the East (Interview 12). To them the privatisation of BWB was part of a process of re-making Berlin; creating a new – neo-liberal – reality in the city. According to a consultant who worked at Finance during the bidding process, some in the government saw privatisation as a means of breaking-up the complacent and corrupt "old boys' network" in Berlin politics (Interview 11), a view supported by the journalist (Interview 4).

This was about more than just changing the way public companies worked. It was also about transforming the perceived culture of governing in Berlin. Privatisation was part of the wider programme of reforming the public sector along the lines of the NPM. It provided a means through which one policy sector – the water sector – could be re-cast. It was a "tool to reconstruct the company (BWB)" (Interview 11). The aim was to utilise the 'know-how' of the private sector to achieve competitiveness in the water sector, as the Union representative stated in an interview (Interview 17). To the Finance Senate this explicitly meant international companies and not merely German-owned companies (Interview 14). To truly change Berlin proven global players were needed.

Securing privatisation as a policy option June 1997-July 1998

Between the formal proposal of privatisation by the Senate (24th June 1997) and their decision to proceed with a partial privatisation of BWB according to the 'Holding Model' (8th July 1998), there was much opposition to privatisation. Action was taken by the Unions within a couple of months. On the 13th August 1997 thousands of BWB employees demonstrated against privatisation (Berliner Zeitung 1997d), arguing it would lead to job losses and higher water prices. In the same month, at the Berlin SPD Party Congress, a vote on BWB privatisation was delayed due to opposition from the Parliamentarians and more particularly the Berlin regional party members (Interview 16). The Opposition Parties – *Die Grünen* and the PDS – were also strongly opposed to any form of privatisation (Interview 7).

Of this opposition, the most important for the Coalition Government in this period was that of the BWB Workers and the SPD. In the short term, the SPD leadership (and consequently the governing CDU-SPD coalition) were not able to approve the privatisation of BWB. Indeed, the divisions within the SPD, at least partially, explain why it took a year for the Senate to be able to formally approve the privatisation of BWB. In fact, the eventual approval came the day after a majority of the still divided SPD voted in favour of partial privatisation.

Assessing this period is difficult. On the one hand, as noted, opposition seemed strong. Moreover, the government was somewhat hesitant, wary of this opposition and willing to compromise. Here the continuing importance of the Unions in politics is apparent. The CDU and the SPD both stated in the summer of 1997 that privatisation would only happen with the agreement of the BWB workers and the Management Board (Berliner Zeitung 1997b). This early statement, though no doubt an attempt to allay the fears of both, would, in the case of the workers, be proven to be at least partially true. Though the government ultimately rejected the Union's alternatives to privatisation, a compromise was sought with them. Furthermore, despite all the speculation, the apparent interest in BWB and statements in the press, a quick privatisation sale did not occur. Instead, the Finance Senate announced their intention to take $600 million (DM1 Billion) from BWB's equity – and not the $1.2 Billion (DM2 Billion) that KPMG had suggested – to service Berlin's debts (Berliner Morgenpost 1997). Formally, this decision did not require the approval of Parliament. Some, such as the civil servant at Finance, have seen it as virtually pre-empting the decision to privatise BWB (Interview 14). This is not necessarily the case as taking equity from BWB emerged as a late alternative to privatisation in 1999, proposed – however sincerely – by the CDU leadership.

At the beginning of 1998, neither the CDU Senator for Economy, Pieroth, nor Fugmann-Heesing at Finance enjoyed the full support of their parties for privatisation (Riecker 1998). Like their parties, both were primarily concerned by the BWB workers threat to strike if privatisation was implemented (Riecker 1998). Still, a certain fatalism amongst opponents is discernable; a sense that despite the opposition, privatisation was inevitable. This period is characterised by the emergence of high-profile lobbying from bidding companies and the presence of consultants working for them, as well as political parties, even the Unions and sometimes more than one of these organisations at the same time. It is likely that lobbying had occurred for some time but the very public nature of it from the summer of 1997 must have reinforced the sense that privatisation was something of a foregone conclusion.

The SPD politician working at Finance stated that the politicians were "schizophrenic" regarding privatisation: outlining their opposition to privatisation in public, whilst implicitly accepting that it would happen (Interview 12). As mentioned, privatisation of BWB was an official policy proposal of the 1995 CDU-SPD Coalition agreement. It had, in principle, been agreed by the party leaderships, if not the Parliamentarians and the broader party memberships. Additionally, this politician stressed that the Parliament's approval of the Senate's budgetary plans for 1998 included $1.1 Billion (DM2 Billion) from the sale of BWB (Interview 12). This provided indirect approval for the sale of BWB for $1.1 Billion.

The SPD's decision to support privatisation came after a period of much internal party debate. Clearly this was to be expected from the centre-left party in Berlin. Despite the shift at the national level to the 'Third Way', mirrored in the Berlin leadership, the bulk of the Berlin SPD was still considered to be on the left according to an SPD parliamentarian (Interview 16). This interviewee stressed that what was seen as the right wing, as represented by the party chairman Böger, Fugmann-Heesing and Klaus Wowereit (then spokesman for Financial Affairs, now Mayor of Berlin) was isolated from, and very unpopular with, much of the party rank-and-file. They did, however, dominate the leadership with the only prominent representative of the centre-left being Senator Peter Strieder (Senate for Construction, City Development, Environment and Transport). An SPD parliamentarian emphasised the divisions within the party at this time, but that Fugmann-Heesing made negotiations very difficult as she "wanted no compromises" regarding privatisation (Interview 16).

On the 15th November 1997, at a special SPD party convention, the 'conversion model' (see Table 4.1) was narrowly rejected (149: 141) but support for some form of privatisation was agreed. By June 6th 1998, at a meeting of the regional SPD Party, a majority voted in favour of partial privatisation following the framework of the 'Holding Model' with the further condition that no more than 25% of the company should be sold to a foreign company (Wetzel 1998). The stipulation of these conditions has been interpreted as a means of appeasing the BWB workers and the Unions (Berliner Hoffbauer 1998). This may have been true, but the complex model itself revealed the continuities in policy-making in the period – it had been used for the Bankgesellschaft. In the Holding Model, BWB legally remained a public entity but its operations – this time including WSS within Berlin – were commercialised (Lanz and Eitner 2005a 4). In this model BWB would became a subsidiary of a Holding *Aktiengesellschaft* (Public Law Company or Plc.), with the majority share of both being owned by Land Berlin.

The 'Holding Model' entailed the foundation of a *Beteiligungsgesellschaft* (Investment Company set up by the private investors), which would buy 49.9% of the shares of a Holding Plc., formally established by the State of Berlin who would remain majority shareholders. These new companies, the *Beteiligungsgesellschaft* and the Holding would then enter into silent partnership agreements, giving the private investors a total of 49.9% control of the public-law entity BWB (*Anstalt öffentlichen Rechts*). The State of Berlin therefore kept a majority share of the Holding and the *Anstalt*. This was the key condition of the Holding Model.

On paper at least this model ensures that the balance of power between the private and public partners (Berlin government) was tipped just in favour of the latter. This compromise deal cleared the way for the Senate to formally announce the proposed privatisation of BWB using this framework on the 8th July 1998. Although the party ultimately supported privatisation in these two votes, the leadership was never entirely certain of what the outcome would be (Interview 16). Though they might not have been certain, they had good grounds to be optimistic that a compromise such as partial privatisation would succeed. When the SPD voted on privatisation of the Gas and Electricity Companies, they had always – albeit marginally – voted in favour. The SPD parliamentarian explained this with reference to the increasing grip of the 'Third Way' during this period and the severity of the fiscal crisis (Interview 16). This created "huge pressure" (Interview 16). Without this, the Berlin SPD, traditionally seen as being on the left of the SPD, would never have voted in favour of privatisation (Interview 16). Put in different terms, this "huge pressure" could be attributed to the potency of the global city policy discourse: the way in which it provided a Berlin-specific means of interpreting the fiscal crisis, of creating the 'need' for privatisation.

In addition to these 'huge pressures', the SPD leadership, Fugmann-Heesing apart, worked hard to find a compromise (Interview 16). Part of this strategy rested on employing a private sector consultant to provide an expert report. A consultant, Franz Joseph Pröpper was asked to write a report on how BWB could be privatised within a framework which allowed for some continued public involvement. Pröpper proposed the Holding Model.

This proposal proved influential. He presented his findings to the Party a week prior to their special convention in June 1998, where they had approved the partial privatisation of BWB according to the model. Problematically and apparently without the knowledge of the SPD (Schomaker 1998), Pröpper had also that year been employed by RWE to do exactly the same job. He had made the same proposal: a partial privatisation based on the Holding Model and additionally outlined a lobbying strategy for securing the purchase of BWB (Interviews: 1, 8; Schomaker 1998).

Given the SPD's consistent, if never unanimous, support for privatisation in the 1990s, the Union representative stated in an interview that he felt that the privatisation of BWB was, from the start, inevitable (Interview 17). The strong pro-privatisation stance taken by the SPD leadership and the passivity of the Parliamentary Party were the grounds for this fatalism (Interview 17). Throughout the privatisation process the SPD leadership, most starkly Fugmann-Heesing and Böger, were often more vociferous in their support for privatisation than the CDU leadership. It could be argued that they were simply more open about their intentions, while the CDU leadership were more circumspect, particularly with an election due in October 1999. A number of interviewees claimed that the CDU played political games – especially in 1999 (Interviews: 9, 12, 15).

Securing the support of the SPD for a partial privatisation removed the major obstacle to privatisation within the formal system of politics. Though there was no absolute guarantee that the Parliamentarians would vote in favour of the final privatisation deal, it had virtually been secured. With the largest party, the CDU, also in favour of privatisation, the support of a majority within the SPD party, traditionally opposed to privatisation, removed the natural base for political opposition. By July 1998, the Unions and workers were left looking increasingly isolated in their opposition, with only the smaller opposition parties and a few civil society groups opposing privatisation.

Unions, BWB workers and the CDU: the guarantee of job security

If the support from the SPD allowed the privatisation process to progress through the formal procedures of the political system, coming to some kind of agreement with the Unions would prove crucial to neutralising opposition more generally. The accommodation of the interests of the Unions and BWB Workers occurred alongside the SPD internal debate, with an agreement between Land Berlin and the BWB workers being signed the day before the formal Senate decision to privatise on the 8th July 1998. The chronology here is significant as it indicates the importance of the Unions and BWB Workers to the Senate: the formal political decision came only after agreement was reached with the Unions and workers. A further agreement between Land Berlin and the BWB workers would be signed on April 19th 1999 (confirming the earlier agreement), again preceding the official parliamentary discussion. Officially, at least, this deal ensured that there would be no job losses at BWB for 15 years (Riecker 1999).

The pressure exerted by the Unions and BWB workers prior to this, most notably the threat to strike if privatisation was approved placed pressure on both the CDU and SPD throughout this period and was seen by the Berliner Morgenpost (Riecker 1998) as the reason the government agreed to the compromise of a partial privatisation. Though one might expect this pressure to have had the most impact upon the SPD leadership and parliamentarians, in Berlin this pressure also applied at least equally to the CDU. This was not simply because they were part of the governing coalition but was also due to the fact that they enjoyed close relations with the Unions and BWB workers. It can be seen as indicative of a corporatist style of governing in which compromises were sought and deals made with Unions prior to any official decision being made.

Serving CDU politicians of the period were generally unwilling to be interviewed, with the exception of the politician working at Finance. However, the closeness of the CDU to the Unions and the BWB workers was stressed by many interviewees (Interviews: 9, 11, 16, 17). In fact, despite the SPD's traditional links to the Unions and workers in Germany, it was the CDU who enjoyed the better relations with them in Berlin. One consultant even claimed that later in the process something of an "alliance" emerged between them (Interview 9). Water management was traditionally one of the CDU policy sectors – a sector in which they had generally taken the lead when in Coalition government (Interview 2). This support from the CDU was by no means unanimous and was often, when expressed by their leadership (Landowsky and Diepgen), designed to wrong foot the SPD (Interviews: 16, 15), particularly in the latter stages of the privatisation process. Nonetheless, the links were strong.

This compromise was engineered by the CDU leadership, particularly the famed political operator Klaus Landowsky, who had one eye on the elections in 1999. He was apparently very charming and persuasive in his dealings with the workers and Union, being able to convince them of his support with his speeches and networking skills (Interview 16). Regardless, of the CDU's intentions, this deal neutralised not only the main argument against privatisation put forward by the Unions, but one of the key arguments proposed by all opponents of privatisation – that jobs would be cut. According to a Left Party politician, the deal secured the BWB workers the best working conditions in Germany (Interview 2). A view supported by Riecker (1999). It also, arguably, contradicted the senior SPD politician's objective of changing the "mentality" of Berlin.

As stated, opposition from the Unions was strong throughout the summer of 1997 and 1998, in public at least. For instance, in September 1997 a lawyer working for the Unions emphasised that privatisation, the creation of a private monopoly, would be unconstitutional (Der Tagesspiegel 1997c). Despite their agreement regarding job security, the Unions also proposed alternatives to privatisation right up until the signing of the deal. In May 1999, for example, the Unions alternative to privatisation was to retain BWB as an AöR and establish a new company, Berlin Wasser, as an AG, to be sold on the Stock Exchange.

However, there is a sense that the Unions were always looking for a compromise, a sense of fatalism about their opposition to the privatisation. The early guarantee from both Landowsky and Böger that no privatisation deal would be implemented without their consent probably indicated that there was room for them to secure their own positions within some form of privatisation. According to a BWB workers representative at the time, there was some acceptance that reform of BWB, was required: a general recognition that BWB was poorly managed and that change was needed (Interview 10).

Going further, the Union representative suggested that though the official line of the Union was opposition to privatisation, there was an acceptance that it was likely to happen from a relatively early stage. There were even small pockets of BWB workers who supported privatisation (Interview: 17). As far as he was concerned, the Union's only interest was securing jobs and good working conditions for the BWB workers in the privatised BWB. As other researchers (Fitch 2007a; Passadakis 2006) have stated, the Unions represented the best means of mobilising opposition to privatisation. Some on the left have criticised the Unions for not trying harder to engage with the public (Interviews: 1, 2). It should, however, be remembered that by the time they signed the deal with Land Berlin, the partial privatisation had been approved by the SPD.

While the opposition to privatisation from the SPD and the Unions can be seen to have secured a partial privatisation, their ultimate support for this compromise made privatisation almost inevitable. The lack of solidarity between the Unions and the SPD limited the potential of both to promote alternatives, to build an opposition. Further it appears that the Unions did not have, nor did they develop, particularly strong links with the main opposition parties: the PDS and the Greens. The strong links between the Unions and the CDU and SPD, the traditional parties of government, appear to have precluded this. Having no real leverage within the parliamentary system and little influence on negotiations with the Unions and workers, the only real option for the opposition parties to broaden support against privatisation was to engage with actors outside of the system of politics: NGOs and civil society groups.

There was, however, little organised activity against the privatisation of BWB, and no successful attempt to mobilise the public. Some small NGOs did campaign against privatisation, such as *Wasser in Bürgerhand* ("Water in Citizens' Hands") and *Attac Germany* (Fitch 2007a: 143). They were not, however, able to make themselves visible in the media and were unsuccessful in more direct campaigning. Fitch's search of the left-wing newspaper Die Tageszeitung as well as the centre-right Die Welt revealed no articles about NGOs or citizen groups campaigning against the BWB privatisation issue during the policy process, 1997-1999 (Fitch 2007a: 145). The NGO *MenschenRechtWasser*, ("Water is a human right"), a Protestant organisation campaigning for water rights, complained that their attempts to engage with people on the street were met largely with indifference (Fitch 2007a: 144). Unlike a similar attempt to engage with citizens in Hamburg, there was to be no grassroots movement against privatisation. While little, if any, popular support for privatisation is apparent, there is no evidence of widespread opposition.

Conclusion: producing a politics of inevitability

It is difficult to assess how far the partial privatisation of BWB was, from the start of the formal policy process, an inevitability. As stated the Berlin government argued that there was 'no alternative' and sought to strategically set the terms of the debate. The Berlin government was in serious debt by 1997. This Chapter has not disputed that the financial situation was severe. Rather the aim was to reveal how neo-liberalism shaped the context in which the BWB partial privatisation came to be seen, by many, as inevitable. It has exposed some of the normative assumptions that informed policy-making in this period, revealing in addition how the negative consequences of such policies contributed to the financial crisis. This was done with reference to the translations of the global city policy discourse which served to "mobilise bias" (Hajer 2002, 62) in water governance. By the time of the official policy process, neo-liberalism had shaped the terms of the debate. A neo-liberal rationality had been embedded through a series of pivotal policy decisions emanating from the aspiration to re-make Berlin as a global city. These decisions transformed the policy-making landscape. The fact that they were not seriously problematised in the BWB privatisation debate is tribute to the extent to which neo-liberal assumptions had become accepted, if not taken for granted. They framed the range of policy options apparently available, even to opponents of privatisation.

The limited room for manoeuvre, apparent in the alternative models put forward by opponents, did not itself make privatisation inevitable, regardless of what the government claimed. There may have been a sense amongst many actors that there were no alternatives to privatisation but it still had to be achieved. There was still some doubt. This was revealed most subtly perhaps in the unclear positions adopted by certain actors. If neo-liberalism was potent in this period, it was not all-powerful. This chapter highlighted the ambivalence of actors: the sense that their positions were not fixed, but in the process of being determined. It was a time of loose positions and unclear interests; of exploring the limits of other actors' interests, making compromises, perhaps even of saying one thing and meaning another. The point here is that a politics of inevitability had to be produced in the policy process. It required the construction of relations (securing agreements of support) and the production of new knowledge (that privatisation was necessary and the model to achieve it).

Following Callon (1986) we could characterise such interplay between actors as a phase of *interessement* in which identities and interests had yet to be "tested"; to be exposed to "trials of strength" between implicated or otherwise interested actors (207). From this perspective, interests and identities are "formed and adjusted only during action" (Callon 1986: 207). Actors were exposed to the strategies of others: the attempts by other actors to define their own interests and identities. In Foucauldian terms, these were unstable power relations, characterised by resistances and "duels" between actors (Dean 2007, 9). This is most clearly seen with regard to the SPD, the Unions and the Workers, whose initial opposition, when exposed to the strategy of accommodation by the government, eventually gives way to support (albeit far from 100%). From the research conducted, there is a sense that even amongst these sources of opposition many were willing to seek a compromise from quite early in the process. It is clear that the Berlin government was also willing to compromise. There was no fixed, shared vision of how BWB should be privatised: the Holding Model was not forced through by the government. With an election due the following year, politicians, perhaps especially the CDU, were keen to realise the privatisation with as much support as possible. The shape of the formal privatisation policy was produced through these trials of strength: constructed through these compromises within the frame that privatisation must happen. In particular the continuing importance of the Unions in politics is apparent in the assurances offered by the government regarding the workers.

Obviously the formal institutional powers of the government were important in this stage of the policy process. The government proposed privatisation, played a calculated role in setting the terms of the debate and had the formal privilege of devising the policy. Once the government had gained the support of its party members and had agreed a compromise deal with the Unions, it was difficult for smaller parties and civil society groups to effectively oppose privatisation. Commenting more generally on privatisation processes, Vollrad Kuhn, former business minister of the *Die Grünen* in Berlin, argued that opposition groups can impede privatisation, but only if they were able to influence parliamentary majority parties through utilising the media and working in alliance with Unions, workers and civil society groups (Fitch 2007a: 138). Kuhn's point here is that, given the increasing acceptance of privatisation amongst the larger political parties, privatisation can only be resisted outside the formal system of politics. Thus though it can be said that those opposing privatisation had to accept the terms of the debate as set by the global policy discourse, the institutional advantages enjoyed by the Berlin government do play a fundamental role in it being formally approved by Parliament.

What this discussion of the process has revealed is the work involved in exercising these formal powers – in making the privatisation inevitable. Such formal powers had to be translated in the making of the policy. Despite the lack of initial support for privatisation, within a year the Berlin government had provisionally secured a broad base of support for privatisation. This had been achieved through a mixture of compromises with key opposition groups, some persuasion and an unwavering argument that there was 'no alternative' to privatisation. Those in opposition were unable to garner enough support for an alternative to privatisation. With the institutional advantages enjoyed by the Governing Coalition against them, there was a lack of united opposition within the political system and an absence of popular opposition outside of it.

Finally, this chapter has also revealed the growing importance of consultants in policy-making. It has been argued that the embedding of neo-liberalism in 1990s Berlin was bound-up with the creation of new forms of governance in which consultants played prominent roles: from the formal institutional arrangements of the Bankgesellschaft to the informal arrangements with urban marketing firms to promote the New Berlin and the informal advisory roles played by consultants in the commercialised BWB. In other words, a neo-liberal rationality was not only apparent in the form policies took, but also in the means through which politics was practiced in 1990s Berlin.

6. From ready-made accounts to a politics in the making account of the privatisation of BWB

Introduction

The previous chapter revealed some of the discursive and material processes through which the partial privatisation of BWB became 'inevitable'. By exploring policy-making in 1990s Berlin with reference to the global city policy discourse, the aim was to get beyond a simplified account of how and why the privatisation decision was made. The purpose of this chapter is to get beyond more official accounts of the BWB privatisation. This time the focus is on how the privatisation, having been agreed, was implemented. The aim is to assess how the political system in Berlin functioned; to understand how a controversial and complicated piece of legislation was passed.

To do this it is necessary to consider and ultimately move beyond two slightly contradictory ready-made accounts of this stage of the policy process. Both accounts, provided mainly by two senior civil servants, suggest that the privatisation was by-and-large implemented through the formal, traditional institutions of the political system. Where they differ is in the role they attribute to hired private sector consultants. Still, they are both in their different ways representative of – official – ways of thinking about government which serve to conceal the actual arrangements of actors and power structures through which the partial privatisation was ultimately realised.

Ready-made politics 1

The first ready-made image of the process is provided by a civil servant at the Senate for City Development, Construction, Environment and Transport, who worked on the environmental aspects of the privatisation. It can be seen as alluding to the classical-modernist (Hajer 2003a) model of policy-making, a Weberian image of government generally still potent in Germany (Schedler and Proeller 2002, 170). It is re-telling which describes policy-making through the traditional democratic model of institutions and formal processes. Indeed, through operating slightly apart from societal actors, the political system somehow rises above particularism and brings a form of objectivity to decision-making. In this account, consultants played a marginal role in decision-making. Implementation of decisions was explained by looking at the institutions and formal processes of democratic systems which were simply robust enough to fairly represent the varied interests of the society the state governs.

Ready-made politics 2

The second ready-made account is provided mainly by the civil servant at Finance and the SPD politician in the Coalition Government. It is more expansive in its discussion of consultants. Suggestive of the thinking embodied in the managerialist view of government, this account emphasises the importance of consultants in dealing with the 'business' of privatisation: the management of the bidding process. Consultants are vital as they supply government with the private sector expertise required in privatisations. Their roles are, however, restricted: they are brought in with a specific remit. In this ready-made version the formal political institutions remain the means through which political decisions are made. What is different in this account is that a space is demarcated for the private sector consultants to work alongside politicians and civil servants. Ultimately, however, the account still makes claim to objectivity and legitimacy, provided through the classical-modernist political system.

The politics in the making account

Pieced together from interviews with a wide range of participants, the politics in the making account provides a more problematic picture of policy-making. The chapter does not claim that this is a definitive account of the process: it is yet another version, albeit one formed from a broader range of sources. Instead, the politics in the making account emphasises the need to problematise and look beyond such models. It underlines the extent to which *a priori* assumptions about governance structures cannot be made: that a polity is not pre-given but is also produced through governance processes. It too needs to be translated. In this less stable view of policy-making, the polity is an outcome as well as an input in governance processes (Hansen and Sorensen 2005, 93); it shapes and is, in turn, shaped by policy-making processes.

Table 6.1. Contrasting accounts of the privatisation process.

ACCOUNT	WHO	IMAGES OF POLICY-MAKING AND POLITY	IMPLICATIONS
READY-MADE 1	Civil servant: Environment.	Classical-modernist. Bureaucratic. Elitist. Consultants not central to process.	'Checks and balances' and independent bureaucracy = indirect representation and accountability. Transparency not important. Participation of other actors not important. Political system worked- it provided legitimacy.
READY-MADE 2	Civil servant: Finance. Senior SPD poltician.	Managerialist government. 'Governance'. Political institutions make decisions, consultants assist in implementation.	Informal, ad hoc arrangements complement formal institutional processes. Clear, limited remit for consultants. Core of classical modernist system remains. System works in changed context of governing.
POLITICS IN THE MAKING	Multi-actor account.	Polity made through policy process. New, extra formal institutional arrangements centred on consultants. 'Duels' with other institutions and actors.	Consultants vital to privatisation in 'Finance team'. Rules and remit unclear. Contests between formal and informal processes. Legitimacy, accountability and transparency problematic. Classical-modernist system did not work as claimed.

The partial privatisation of BWB is described in terms of the emergence of a new structure of governance which encompasses parts of the traditional political system, but aims to reform it and break free from some of its constraints. This informal, ad hoc set of arrangements centres on the team of consultants employed by the Finance Senate to manage the privatisation (the 'privatisation management team'). It can be seen as another example of the public-private forms of governance characteristic of policy-making in the 1990s. The emergence of this public-private management team contests the formal institutional rules and processes of policy-making: it re-orders, however temporarily, the polity itself. It can be interpreted as an attempt to rearrange relationships of power; one which places consultants at the heart of governmental decision-making. It is, however, resisted by other actors, including some governmental actors who, though formally in favour of privatisation, feel their positions and interests are being undermined. As a result, power is never fully possessed and exercised by one set of actors in the policy process. Rather it is defined through the exchanges between them: it is the outcome of their relationships (Dean 2007, 9).

July 1998- October 1999: realising the privatisation project

The formal announcement that the Senate would proceed with a partial privatisation of BWB on the 8th July 1998 could, in conventional terms, be seen as the implementation stage of the policy process. There was still much work to be done and the potential for problems to arise remained strong. The legal framework for privatisation had to be finalised and formally approved first by the Senate and then by the Parliament. This was complicated because there was no legal basis for privatisation of water services in Germany. Additionally, the exact objectives of the Berlin government in the privatisation deal were still to be determined. For example, apart from the sale price, what should be the priorities of negotiations with the private investors? How were the environmental, technical and social aspects of privatisation to be dealt with? How was the sale to be realised – what strategy should be employed to deal with the private companies? The Holding Model may have been the agreed means of organising the partial privatisation, of ensuring democratic control of the public-private company, but it remained a mere model, the exact details of which had to be legally, legislatively and commercially refined.
Some opposition was still apparent during this period. The Union and the BWB workers were still – officially at least – threatening to strike against any privatisation. Potentially, this could disrupt the task of finding a buyer for BWB (and the objective of securing a high sale price). Additionally, the roughly "3 million water experts" (Interview 9), as a consultant for Finance called the Berlin public, were another obvious concern for the supporters of privatisation. Taken together, the legislative, legal, commercial and public dimensions of the privatisation process created the potential for much complexity and controversy. For these reasons, to think of the period of July 1998 – October 1999 as one merely of 'implementation', as the ready-made accounts suggest, is misleading. The extent to which policy

definition was still ongoing is suggested by the very length of this period. The initial plan was for the privatisation process to be completed in March 1999 with a sale price of at least $1 Billion (DM2 Billion) as the target (Berliner Zeitung 1998).

The establishment of the Steering Committee and the team of consultants at Finance (July 1998-October 1998)

With the announcement that they would proceed with a partial privatisation on the 7th July 1998, the Berlin Senate established a "Steering Committee" to oversee the partial privatisation of BWB. This consisted of the Senators for Economics, Finance and City Development, Construction, Environment and Transport[29]. The Senators were: Elmar Pieroth (Economy: CDU), until November 1998, then Wolfgang Branoner (CDU), Annette Fugmann-Heesing (Finance: SPD) and Peter Strieder (City Development). Alongside the Steering Committee were the Parliamentary Party Leaders of the CDU, Landowsky, and the SPD, Böger. Constitutionally, the Senator for Economics had overall responsibility for publicly owned companies such as BWB, while the Finance Senator was responsible for fiscal management and public property. As in the privatisations of the Electricity (BEWAG) and Gas (GASAG) companies, Fugmann-Heesing was handed the task of overseeing the bidding process. The Senate for City Development was responsible for the environmental aspects of water management and was included to ensure technical and environmental aspects were duly considered. Between them the Senators were supposed to represent the range of public interests in BWB.

Alongside these political institutional arrangements, Fugmann-Heesing assembled, at great cost, a team of consultants to manage the bid in the late summer and autumn of 1998. Contained within the Finance Senate's team were consultants from three companies. Merrill Lynch, the investment bank and financial consultancy, acted as financial advisors and were overall leaders of the bidding process (tasks they had performed for the privatisations of BEWAG and GASAG). Hengeler Mueller, a large German law firm, acted as legal advisors and the tax consultancy, BDO (Deutsche Warentreuhand AG), advised on the valuation of BWB and the setting of water tariffs. It is also worth noting that the Senate for Economy retained the services of the law firm White & Case to work internally, particularly on the Partial Privatisation Law (*Berliner Betriebe Gesetz, TPrG*), though they did not officially participate in negotiations with the bidding companies.

The premise for forming this privatisation management team was to utilise their specialist knowledge of these areas and to ensure that the Berlin government

[29] Senate for City Development from now on.

was able to deal effectively with the private sector companies interested in purchasing BWB (Interview 12). Ultimately, this move can be seen as an attempt to 'depoliticise' the bidding process. Consultants were officially meant to be the only real contact between the Berlin government and the bidding companies (Interviews: 9, 14). They were meant to provide a buffer between the political system and the commercial process of privatisation.

Officially at least there appeared to be a division of duties and responsibilities. The Steering Committee had overall control of the privatisation: it was meant to decide upon the various implications of any privatisation deal. The team of consultants at Finance managed the bidding process, negotiating on behalf of Land Berlin. This was the view presented by the two civil servants who worked on the privatisation. The interview with the civil servant at City Development was conducted towards the beginning of the fieldwork, prior to interviews with consultants. In the interview the civil servant was challenged to clarify his own role, the roles of the other key institutional actors and the process more generally. Questions were asked about the consultants but without reference to specific tasks they were alleged to have performed. By the time of the interview with the civil servant at Finance, interviews with some of the consultants had been carried out. It was therefore possible to challenge his account with specific reference to their version of events. During this process the civil servant revised his initial description of the process.

In both of these accounts, something is learnt about the privatisation. The argument put forward is not that the ready-made accounts should be seen as entirely misleading. Rather it is asserted that the ready-made accounts should be interpreted as constructs which, in drawing on models of the political decision-making system, provide a clear, relatively unproblematic image of the privatisation process. Crucial, problematic elements of the process, which emerge from interviews with other actors, are not apparent in the ready-made accounts. Obviously, this may, in part, have been intentional. The BWB privatisation is still a source of controversy and those directly involved probably felt the need to defend their roles and the process more generally when interviewed. Additionally, however, it should be accepted that few, if any actors had an overview of the process. The purpose of the politics in the making account is to bring a range of perspectives together to produce a fuller, if more complex account of the process. The politics in the making account is, therefore, one more account of events. However, it is argued that a forceful enough picture of a problematic policy process emerges for there to be concerns over the way in which the privatisation was conducted.

It is here that a more "realist" concern – as Miller and Rose (2008, 57) might describe it – for explanation and causation becomes crucial to the assessment of the legitimacy, transparency and accountability of the decision-making process. A detailed look at this stage of the privatisation process shows that much hinges on the sequence of key events. As shown below in Table 5.2, alongside the agreements with the BWB workers (7th July 1998 and 13th April 1999) two key processes can be discerned in the privatisation of BWB. First, there is the formal parliamentary and legal process through which the Partial Privatisation Law was eventually agreed. It was proposed by the government, discussed and approved by the Parliament between January and 29th April 1999. It was, then, almost immediately challenged in the Constitutional Court by the Opposition Parties (May 1999). Finally, on the 29th October 1999 the Court ruled that the law was legitimate with some exceptions, most notably the profit calculation formula, 'R +2' (revenue + 2%).

The Partial Privatisation Law aimed to implement the Holding Model, through legally enabling BWB to enter into negotiations with non-public entities about partnerships. The Holding Model aimed to ensure that Land Berlin had overall control of decision-making in BWB in the public-private partnership. This was the compromise agreed in the formal policy debate between June 1997 and July 1998. The Parliamentary and legal processes are the traditional, formal process through which policy was made in Berlin. Alongside this, the 'bidding process' deals with what were termed the 'commercial' aspects of the privatisation: finding a buyer, devising contracts and agreeing a sale price. This process was officially managed by the hired private sector consultants and resulted in both public and confidential contracts with the private sector companies. If timing is important to the assessment of this process, so is the extent to which these political, legal and commercial processes – and the personnel operating within them – do or do not overlap.

Ready-made politics 1

The civil servant at the Senate for City Development accepted that the entire implementation process, and not merely the bidding stage and the negotiation of contracts with the private partners, was neither very transparent nor inclusive. Rather it was contained within the executive-led bureaucratic system centred on the Steering Committee. This was not seen as a problem: the system worked because of its expertise and because it is, by design, roughly representative of the public interest: the Senates represented the various – financial, economic, environmental and technical – dimensions of the public interest. In this view of the policy process, politicians and administrators are generally bound by the responsibilities of office and, even if not, they were prevented from distorting or corrupting the process by the broader institutional rules and structures within which they worked (Interview 5).

The civil servant thus provides a classical image of a competitive policy process in which the three Senates, despite pursing their own interests, are ultimately forced into compromises. These compromises represent the effective operation of the bureaucratic system. The institutional decision-making system is thus roughly objective – it rises above personal and political interests. In this view the implementation of policy was complicated but rational and relatively linear (Interview 5). It was stressed that the process was thorough and democratic. It was democratic because the Berlin government had the support of the majority of the Parliamentarians (Interview 5). The government had a mandate to carry out the privatisation. It did this by establishing the Steering Committee and putting the city's administration, with its rules and set procedures designed to ensure legitimacy and representation of interests, into action (Interview 5). The administration, with guidance from the Steering Committee then engaged with the bidding companies to thrash out a deal. Once reached, the Parliament debated and eventually approved the deal (Interview 5)

Nonetheless, the Senates for Economy and Finance, led the process and tried to make it as "smooth" as possible for the potential buyers (Interview 5). Their priority was to achieve a high sale price. As such this meant that they were – along with the bidding consortiums – unhappy with the technical and environmental regulations put forward by the Senate for City Development (Interview 5). The Economy and Finance saw these measures, which were designed to safeguard standards post-privatisation, as imposing costs (Interview 5). According to the civil servant, the Senate for City Development effectively "disturbed" the privatisation process (Interview 5). They were able to do this because of their formal responsibility for regulating WSS services and the determined attitude of the Senator in charge, Strieder.

The department's initiatives were, it was claimed, 90-95% successful (Interview 5). Their influence can be seen in two important measures in the final version of the Partial Privatisation Law (Interview 5). First, the buyer must use Berlin for sourcing and providing water, though this was in fact already a part of Berlin's water law. This measure is aimed at encouraging efficiency and investment in infrastructure within Berlin. Second, a municipal department regulating groundwater levels was established in Berlin. According to the civil servant, this department has ensured, against BWB's wishes, that the smaller waterworks remain open (Interview 5).

Table 6.2: Key events July 1998 –October 1999

7 July 1998: Agreement on jobs between Senate and BWB workers. Leads to:		
8 July 1998: Berlin government's official decision to privatise BWB according to the Holding Model. 'Steering Committee' established.		
<u>Parliamentary/ legal process</u>	**<u>'Bidding process'</u>**	**<u>Workers/ Unions</u>**
	August 1998: Merrill Lynch appointed September 1998: Hengeler Mueller and BDO appointed October 1998: Bidding process opened	
8 January 1999: Parliament presented with Privatisation Law 14 January: 1st reading of Law		
	Mid-February 1999: Confidentiality agreement with bidding companies March 1999: Final short list of bidders	
29 April 1999: 2nd reading: approved (106-67) May 1999: Opposition parties – launch legal challenge.		April-May 1999: Formal jobs agreement & Unions formally accept privatisation
1July 1999: Parliament approves consortium contract	June 18 1999: Consortium contract: RWE/ Vivendi/ Allianz win July 1999: Confidential amendment: ensure conditions of 'R+2'	
11 October 1999: Berlin Elections 21 October 1999: Court rules 'R+2' unconstitutional	21-29 October: Berlin-buyers negotiations	
29 October 1999: Parliament approves privatisation	29 October: BWB formally partially privatised	

Overall, the civil servant argued that very little changed with the privatisation deal and that environmental issues remained well protected. National German Law states that wastewater systems must be publicly controlled and thus the Berlin government alone – theoretically at least – sets regulation for sewerage treatment every 4 years. Thus, in his view, the partially privatised BWB only really controls supply and treatment as well as the subsidiary companies (Interview 5).

Assessing ready-made politics 1

Strieder and the Senate for City Development have been credited with managing to incorporate some environmental safeguards into the final agreements with RWE/ Vivendi/ Allianz (Lanz & Eitner 2005b). A consultant working for Finance confirmed in an interview that there was a lot of conflict between Finance and City Development, though he attributed this as much to SPD internal politics as scrutiny of the Partial Privatisation Law and contracts (Interview 9).[30] In this sense the ready-made 1 account of a bureaucratic system of checks and balances is, to some extent, convincing. Nonetheless this version of events is problematic. If we take the account on its own terms and accept the view that the policy process does not have to be inclusive of a wide range of actors to be representative of the best interests of Berlin, then the focus turns to how effectively the formal political institutions functioned.

The assumption is that the Steering Committee provided the necessary checks and balances. However, even the civil servant stated that control of the process was dominated by the Senates of Economy and Finance (Interview 5). It was not a decision-making process in which all three Senates were able to exert equal influence: it was a process led by Economy and Finance. It was not therefore equally representative of the range of concerns prompted by privatisation. It can now be said that Berlin's interests were defined, first and foremost, as being (unsurprisingly) about the achievement of the sale price. This prioritisation may not have had led to a disregard for environmental issues, but it does raise questions over the ability of the Steering Committee to represent the wider interests of the city.

Actors outside the government confirm this prioritisation and the effects it had on decision-making; on who participated and what was deemed to be important. The BWB management, despite their intimate knowledge of the company and general support for privatisation, were effectively marginalised. This was achieved through the Steering Committee's early decision to allow BWB's

[30] Strieder was a far more senior figure within the Berlin SPD and represented the left wing (Böger and Fugmann-Heesing the right). Apparently, Strieder tried to exert his authority in the process but Fugmann-Heesing response was "resolute, not diplomatic" and this resulted in a "lot of friction" (Interview: 9).

management (12 Directors) only one voice in the bidding negotiations. This meant that only Wieczorek, the Chairmen, could communicate their concerns and queries (Interview 3). As a result, a member of the BWB Management Board at the time claims that no one from the Senate or the Parliament asked about the implications of privatisation for the infrastructure network (Interview 3). The role of the Management Board amounted to little more than "sitting watching bidders' presentations", providing the private companies and those in charge of the process with the documents they needed for negotiations (Interview 3). This is important because the member of the Management Board – a supporter of privatisation – stated that "it was clear any privatisation would lead to cost-cutting" (Interview 3) (and see Lanz and Eitner (2005a) and Hüesker (2011) for discussions of the issue).

It can, therefore, be argued that through the management of the bidding process, and the importance attributed to financial factors, other considerations, such as the environmental and technical implications of privatisation, were not fully considered and the actors representing them somewhat excluded. The final deal with RWE/ Vivendi/ Allianz has been described as placing too much emphasis on cost-cutting, with too little in the contract about quality of water management (Interview 3). The member of the BWB Management Board unreservedly blames the politicians for being too concerned with the sale price and not seeing the big picture (Interview 3). The final deal was described as good for the politicians (who got the money) and the buyers (who got a good deal), but not good for water management (Interview 3). Though the civil servant at City Development argued the opposite, the Union representative also stated that there was very little discussion about environmental and technical management (Interview 17). As soon as assurances about levels of investment had been written into the contracts, discussions apparently stopped (Interview 17). In other words, the money was seen – mistakenly, the Union representative now accepts – as enough of a guarantee that standards would be maintained.

The ready-made 1 account emphasises the formal institutional processes of policy-making. The commercial, bidding process was not seen as important to decision-making. Consequently, the private sector consultants working for the Finance Senate and Economy Senate were seen as having a minor role. The civil servant stressed that their strategic papers and expert reports were not influential in shaping decisions and that the consultants hired by the government had only a marginal role in the implementation itself (Interview 5). This seems questionable, given that some researchers claim that thirteen consultancies were employed by the Berlin government during the privatisation process at a cost of around DM 80 million (Lanz & Eitner 2005a, 9).

In describing the process with reference only to the formal institutional processes and the rules and rationale on which they are based, this account attempts to simplify the privatisation. It is in this sense that the account is ready-made: a simple potted history, packaged and presented within the formal decision-making processes of the Berlin government. It could be interpreted as part of a rationalisation after the event, comparable to the forgotten controversies of scientific experimentation to which Latour (1987) refers. Politicians and administrators are also inclined to overlook the uncertainties of policy-making, to gloss over the processes through which 'facts' were produced and decisions made. The stakes for them are arguably higher, given that their decision-making aspires not only to rationality and objectivity but also to democratic legitimacy.

It is in this context that models of political systems and, vitally, the claims which they make, can become useful props to describe how policy is made. Certainly, allusions to models of political systems are apparent in the two ready-made accounts. These models provide a neat, black box within which to place the privatisation process. They are potent because they are, at least partially, convincing: they outline discernable processes in policy-making. In the first ready-made account, the controversy of the BWB privatisation disappears behind the classical modernist model of democratic decision-making. It is an account which presents the model as a means of ensuring legitimate, objective and rational policy-making. Ultimately, this first ready-made, controversy-free account can be interpreted as a reification of this model. The second ready-made account also emphasises the traditional institutional processes of decision-making, but within a context of 'governance'.

Ready-made politics 2

A broader account of how the privatisation was realised is provided by the civil servant at Finance and the SPD politician in the Coalition Government. Both emphasised the importance of the consultants they worked with. They placed the consultants working for Berlin to the fore in one aspect of the implementation process: managing the bidding process. The civil servant presented an image of them facing off against the consultants working for the bidding companies around the negotiating table (Interview 14). In this account, the consultants were brought in to 'manage' the commercial aspects of the process. Their remit was clearly defined and limited to the commercial requirements of the privatisation. They negotiated the small details of the final contracts and led the discussions with the bidders (Interview 14). This was for two reasons: politicians were not permitted to deal directly with the bidding companies and consultants have experience of brokering such deals. In this view, the consultants helped de-politicise the bidding process (Interview 14). They were a form of intermediary between the Berlin government and the private sector companies. The SPD politician emphasised how important the consultants were for the professional management of the bid: to achieve a high sale price and ensure good business practises (Interview 12).

A separation of the political from the commercial processes

In this ready-made 2 account there is a clear demarcation between the consultants' roles and responsibilities and those of the Berlin government; between the 'political' decision-making process and the 'commercial' necessities of completing the deal. The civil servant emphasised that the government had, by the bidding stage of the process, already set out the key characteristics of the Privatisation Law (Interview 14). One of the consultants working for Finance also argued this, stating that the "parameters (of the privatisation) were already determined" by the time they were hired (Interview 9).

The timing of events and the division between the bidding, political and legal processes are essential to the claims made in this ready-made account. As shown in Table 6.1, the decision to privatise BWB using the Holding Model was formally taken by the Senate on the 8th of July 1998, at which time a Holding Company was also officially established. The team of consultants was gradually assembled from this point on until the opening of the bidding process in October 1998. The Partial Privatisation Law (based on the Holding Model) was devised by the Berlin government between July 1998 and January 1999, when it was presented to the Parliament. Thus the consultants were working for the Berlin government during the drafting of the Partial Privatisation Law. They were not, however, working *on* the Partial Privatisation Law, the civil servant argued, because by the time they became involved all the key features of the law were in place (Interview 14).

The argument was that the Holding Model provided the framework for the Partial Privatisation Law and thus the form the privatisation would take was decided by the political institutions in advance of the arrival of the consultants (Interview 14). Vitally, the civil servant initially stated that the most controversial feature of the Partial Privatisation Law, the 'R+2' formula, was also in place prior to the arrival of the consultants (Interview 14). 'R+2' was a means of calculating profit for the public and private partners: 'revenue plus 2%'. 'R' is calculated in relation to the average percentage revenue made from 10 year, low-risk Stock Market Bonds over the previous 20 years. 2% would then be added on to determine the profit rate for both the public and private owners. This formula was challenged by the opposition parties in May 1999 and ultimately ruled to be unconstitutional by the Court in October 1999.

The civil servant thus draws a clear line between the political and legal decision-making processes and the bidding process (Interview 14). The consultants were important only to the management of the latter. By implication then, this account also excludes the bidding companies from the key political decision-making phase of the process, as they were only officially involved from the start of the bidding phase in October 1998, again, by which time the key features of the deal had been agreed. These are important statements, all of which will be challenged by other accounts.

Ready-made politics 1 was rooted in the traditional institutional processes of representative democracies. Ready-made politics 2 alludes to a slightly different image of policy-making: one inspired by the "new managerialism" (Saint-Martin 2000). In this version, the private sector consultants provide an expertise necessary to the privatisation process. Their knowledge of the private sector is essential to the effective and professional management of the bidding process. They are, however, managers and not makers of policy. The more general political process remains in the hands of the politicians and civil servants.

However, this managerialist account of the process is also problematic, containing a more subtle attempt to marginalise the role played by the unelected consultants. In fact, the account is contested by the consultants themselves, who revealed that they played a much greater role in devising the final arrangements of the partial privatisation. Further, the extent to which the consultants role went beyond the limited remit of managing the 'details' of the bidding process is made clear by other actors. As will be shown, when presented with this information, the civil servant at Finance revised the first account provided.

A politics in the making account

According to one consultant who worked for Finance, the key to managing the bidding process was to keep "it like an auction" for as long as possible; to maintain competition between bidders to ensure a high price (Interview 9). The interest in BWB was huge. It was the key privatisation in the European water market according to the Director of Severn Trent (Der Tagesspiegel, 1999a). As table 6.3 reveals the two largest water companies in the world, Vivendi and Suez Lyonniase des Eaux were involved in bidding consortiums. In an interview, a consultant working for the RWE/ Vivendi/ Allianz consortium even described it as "the privatisation worldwide" during the late 1990s (Interview 8). In part, this was because the company had over 3 Million paying customers. In part, again it was to do with what Berlin was perceived to represent. It was, of course, a prestigious contract, in the newest capital city in the biggest country in the EU. But it was about more than that – the promise of what was to come. The terms of the global city discourse were apparent, particularly the notion that Berlin was strategically placed to take advantage of the emerging markets of Central and Eastern Europe. Vivendi

President Jean-Marie Messier's stated that Berlin was an ideal base for the company to achieve its objective of expanding into German and Eastern Europe markets (Frankfurter Allgemeine Zeitung 1999a). Such an ambition did, of course, match the Berlin government's vision for BWB. In January 1999 Fugmann-Heesing made it publicly clear that her preference was for international – as opposed to German – companies (Frankfurter Allgemeine Zeitung 1999b).

The Senate's main objective, of course, was securing a high sale price. This was increasingly complicated by the performance of BWB's subsidiary companies particularly *Schwarze Pumpe (*SVZ). Still, the importance of BWB could be discerned in the high, "strategic price" RWE and Vivendi were apparently still willing to pay (Rie cker 1999a). Press rumours from the beginning of 1999 onwards suggested that the *Wasserpoker* was going to ensure a high sale price; such was the perceived importance of BWB and the rivalry between the bidders. Around this time Vivendi and Suez had around a 70% share of the global private water market (Swyngedouw 2003, 9). A big, high profile contract such as BWB was of undoubted significance to them both. In February, for example, the Riecker (1999a) claimed that RWE-Vivendi had made an indicative bid of $1.3 Billion (DM 2.3 Billion) and that they had taken over from the Suez-led bid as favourites. The consultant for the RWE/ Vivendi/Allianz bid argued that the significance of BWB and the rivalry between the two giant French water companies explains the relatively high sale price (Interview 8).

The first short list in the bidding process (Table 6.3) revealed that US energy company, Enron, bidding through its subsidiary Azurix, was the main rival to the two French companies. Like Vivendi, who were in partnership with German utility and energy company RWE, Suez also had a German partner, manufacturing and technology company, Thyyssen/ Krupp (replaced by Mannesmann Arcor in March 1999). The fourth consortium consisted of the English water company, Severn Trent along with BEWAG (the Berlin based and partly privatised electricity company) and PreussenElektra (the German energy company), who later withdrew.

Containing controversy: consultants and the 'management' of the bidding process

As was mentioned in Chapters 1 and 2, privatisation of WSS services is rarely, if ever a popular policy. The consultant for the RWE/ Vivendi/ Allianz consortium said water is always an "emotional" issue and "difficult" to privatise (Interview 8). Given this context, the leadership of Fugmann-Heesing was important. She was a "very strong women"; "it (the privatisation) would not have happened without her" (Interview 8). She was "like a panzer tank!" (Interview 14). As important as she may have been, she did not dominate the process nor was she the sole driving force behind the realisation of the privatisation. In fact, the consultants made it clear that she was not involved in the privatisation on a day-to-day basis (Interviews: 9, 11).

Table 6.3: Main bidding consortiums

Main bidding consortiums
RWE and Vivendi: joined by Allianz at the beginning of June 1999 to help finance the bid.
Suez Lyonniase des Eaux and Thyyssen/ Krupp (replaced by Mannesmann Arcor in March 1999 who saw Berlin as a "platform for the Central and Eastern European markets" (Berliner Zeitung 24 March1999). Bankgesellschaft Berlin join consortia in May 1999.
Severn Trent, BEWAG and Gelsenwasser subsidiary PreussenElektra (who dropped out in mid-March 1999)
Enron subsidiary Azurix and later HeLaBa (Hessische Landesbank)

Source: adapted from Riecker 1999a

Due to the controversial nature of privatisation in the WSS sector, the team of consultants hired by Finance had to be on "constant guard" and much of their work, despite the fact they were financial, legal and tax specialists, was "public relations" (Interview 9). As one put it, the proposed privatisation of BWB immediately created "3 million water experts in Berlin" (Interview 9). Though the privatisation process was similar in structure to those of private-private deals, it was much more challenging due to public interest which meant they "always felt under the microscope" (Interview 9). Despite extensive experience of working on privatisations, one consultant stated that the BWB process was particularly tense and eventful. "I could write a book about the Berlin water privatisation…so many times people on all sides said unbelievable things" (Interview 9).

It is important to stress that from the opening of the bidding process in October 1998 until the official completion of the privatisation on the 29th October 1999, debates continued over the model of privatisation and the very decision itself was still being challenged on a number of fronts. As a consultant for the successful bid said, "right up until the end, everything was in the balance" (Interview 8). The Unions and BWB Workers were still threatening to strike up until the signing of the second agreement with the Berlin government on 13th April 1999. In May 1999 the Partial Privatisation Law was challenged by the opposition parties (PDS and *Die Grünen*), which even led to Fugmann-Heesing threatening to freeze the city budget if the law was not implemented (Interview 7).

To avoid such controversy, the privatisation management team attempted to keep information about the bidding process as contained and secretive as possible (Interview 9). The justification for this was that it was in Berlin's interests – that everything was necessary to achieve a high sale price (Interviews: 9, 11). The implication was that everything should be subordinated to this goal – that the bidding process, indeed the broader privatisation process should be defined in commercial terms. This management team thus contested the official powers of the Steering Committee and the wider political system. As the following sections describe, they employed a deliberate strategy of excluding – as far as possible – other actors from negotiations with the bidding companies. The most obvious example of the management team's attempt to cordon-off negotiations came in February 1999 and prompted a backlash from the CDU leadership.

With the announcement of the first short list (see Table 6.3), Finance signed a confidentiality agreement with the bidding companies in mid-February (Schulte 1999b). This aimed to ensure that there would be no public discussion of the bidding process: that everything discussed between Finance and the bidders would remain secret – even from the rest of the Berlin government. The Greens immediately complained that it was an affront to Parliament (Schulte 1999b). It was, however, the 'norm' for privatisation deals (Interviews: 9, 11). Furthermore, this step was only the most official of a more general process through which the management team at Finance attempted to restrict participation in decision-making.

In other words, the Steering Committee may have been the official institution controlling the process, but the Finance Team appears to have had effective day-to-day control of key aspects of the process (Interview 15). Though the Steering Committee may have been included in the overall strategy, they were certainly excluded from the negotiations with the private sector companies over the contracts and the bids. Indeed, institutional processes were seemingly redefined, with the exclusion of many actors from negotiations with the bidding companies, accompanied by the control of information emanating from these negotiations.

From interviews, it is possible to discern the exclusion of prominent figures at Finance; the marginalisation of the rest of the Berlin government, including the Steering Committee; the exclusion and 'management' of Parliament; and finally, the 'management' of societal interests, the Unions, the media and public opinion. This was a major source of tension and a constant complaint from the Senate for Economy but also other actors such as the Unions, the opposition parties and SPD Parliamentarians (Interviews: 7, 10, 16, 17).

As will be shown, the official institutional arrangements put in place to oversee the privatisation were to a great extent undermined by the Finance Team. There was a cleavage between the traditional political institutions and this purpose-built, expert-dominated public-private sector management team. Though elements of the political system were included in the management team (and Fugmann-Heesing had overall control), private sector consultants were empowered at the expense of some political and bureaucratic actors. In effect an extra-formal decision-making process, centred on negotiations between the management team and the bidding companies, emerged within the political process.

Within this extra-formal process, crucial aspects of not only the privatisation contracts, but also the Partial Privatisation Law were discussed and often decided. This attempt to make decisions apart from the rest of the political system was not entirely successful. Crucial parts of the final law and contractual arrangements were influenced by actors in the wider process. Overall, power did not reside with one set of actors – not even the privatisation management team and the bidding companies. The exercise of political power was more contingent and unstable. Drawing on a variety of resources (the formal procedures of the political system, the official powers of political office and strategic use of the media) other actors were able at various points to exert an influence on the policy process.

Formal versus extra-formal processes

Within the Finance Senate

The status of the consultants within Finance was illustrated by a civil servant working in the media and PR department at the time. This interviewee argued that the only information which came out of the management team during the entire bidding process was that regarding the indicative bids and the agreed sale price and this was directed only to the Mayor (Interview 18). Within the SPD, only Fugmann-Heesing's main allies, Klaus Wowereit (Parliamentary Party Spokesman for Financial Policy) and Böger (Head of the Parliamentary Party) were identified as being involved in the key decisions of the period (Interview 18). They were not, however, greatly involved in the general, day-to-day discussions during the bidding process (Interview 18).

The development of an "enclosure" (Rose and Miller 1992, 188) at Finance was confirmed by a CDU politician at Finance. Fugmann-Heesing had overall responsibility for the process, but she left the "day-to-day" responsibility for implementation to the consultants and the civil servants supporting them (Interview 15). Underlining the extent to which a new way of making decisions emerged, the

CDU politician stated that there were "some meetings with only one or two people, often Fugmann-Heesing and the consultants" (Interview 15).

The exclusion of the civil servants in the Media department was underscored by the consultants taking control of the media strategy (Interview 18). The consultants confirmed this. Their work involved "giving advice" to Fugmann-Heesing on "when to speak and when not to, and who should speak" (Interview 9). It seems the result of the strategy was not to speak very often and not merely to the media, but also to politicians and civil servants outside of Finance (Interview 18). The civil servant in the media department claims to have known so little that they had to ask Fugmann-Heesing if a newspaper report bore any relevance to the truth (Interview 18). It was a deliberate policy, this interviewee thought: by not knowing anything, nothing could be revealed to the press or anyone else (Interview 18). In fact, the civil servant claims he had no direct contact with the consultants throughout the process (Interview 18). His only contacts were through the formal channels of the Senate, to Fugmann-Heesing, for example.

Within the Berlin government

In terms of how the institutions of government functioned during this process, in direct contradiction to the civil servant in City Development, one of the consultants argued that the 'Steering Committee' was not that important (Interview 11). Instead, Finance controlled the process through the close cooperation between the team of consultants and the political/administrative side (Interview 11). This view was supported, with qualifications, by the SPD parliamentarian interviewed. The chairmen of the Parliamentary Parties (CDU, Landowsky and SPD, Böger), along with the Mayor were also party to the major, strategic decisions, if not the everyday management of the process (Interview 16). Below them the Senators of Economy (Branoner) and Construction, City Development, Environment and Transport (Strieder, SPD) were much less involved even in the major strategic decisions, according to the SPD parliamentarian (Interview 16). Indeed, there was apparently tension between Finance and the other two Senates included in the Steering Committee (Interview 16).

As discussed above, the civil servant at City Development also stressed this. Strieder represented the left-of-centre faction within the SPD and apparently pressed for protection of environmental standards. One consultant at Finance saw it differently: Strieder "agreed to everything" and he and his Senate had only a "soft role" (Interview 11). Though no other interviewees stated that Strieder's impact was so negligible, another consultant at Finance also stated that there were personal tensions between the two SPD senators adding that Fugmann-Heesing was "resolute, not diplomatic" in her dealings with him and his Senate (Interview 9).

CDU anger at their exclusion from decision-making appears to have increased with the signing of the confidentiality agreement in February 1999. In particular, the selection of the three bidding consortiums for the final short-list on the 14th March appears to have been made with little consultation of the CDU. In March 1999, the CDU leadership suddenly started to suggest that the privatisation should be suspended and alternatives explored. Landowsky, Diepgen and Branoner proposed taking around $1 Billion (DM2 Billion) in equity from BWB (Berliner Morgenpost 1999a). Landowsky also stated that the privatisation was being rushed and must be re-considered (Berliner Morgenpost 1999a) – despite the fact that it had been debated for almost two years by this stage. Now, he and Diepgen stated they had concerns that the Holding Model would lead to a lack of governmental control in a partially privatised BWB (Der Tagesspiegel 1999b). Furthermore, they even began questioning Fugmann-Heesing's competence to lead the privatisation (Riecker 1999b). These direct criticisms of Finance and the privatisation itself coincided with CDU complaints about the lack of information flow given to the CDU-controlled Economy Senate (Riecker 1999b). Confirming the accounts given by many interviewees, the Economy Senator stated that they had no idea about the amount of money being offered for BWB or the more general contents of the bids (Riecker 1999b).

Though undoubtedly annoyed at being excluded, there was also an element of "coalition politics", as the SPD parliamentarian described it in interview, especially with an election due on the 11th October 1999 (Interview 16). One of the consultants working at Finance stated that the CDU were "essentially against it (privatisation) because it was a totally unpopular measure" (Interview 9). The CDU politician within Finance stated that the CDU were often briefing the press against the SPD (Interview 15). Pieroth and then later Branoner (as Senators for Economy) supported privatisation in principle but "played politics – a PR game with the public" (Interview 15). With Finance taking the lead, it was easy for them to shift blame and brief against the deal. As Bolllmann (1999) observed in May 1999, it did seem strange that much earlier in the debate the CDU had considered full privatisation of BWB and then out of the blue in March 1999 they started opposing any form of privatisation.

In this period it is possible to argue that the Finance Team started to lose the 'PR game'. More fundamentally the limits on their inability to 'manage' the process – and define it purely in commercial terms – were exposed. The consultants themselves admitted in interviews that the project could have been better presented (Interviews: 9, 11). This may have been true, but it could also be argued that the very presence of consultants in the policy process and the attempt to control and depoliticise was bound to provoke opposition. Both were inherently problematic.

Consultants and conflicts of interest

Consultants were prominent throughout the privatisation process and more generally in Berlin politics during the 1990s. When, in the summer of 1998, it was revealed that the consultant Pröpper had worked for both the SPD and RWE on the BWB privatisation, the Finance expert for the The Greens stated that "all over the city there are lobbyists in the most diverse clothes" (Schomaker 1998). There were even rumours in mid-1998, that the former US ambassador to Germany, Richard Holbrooke, was discreetly representing an American firm – probably Enron – said to be interested in BWB at this time (Der Tagesspiegel 1998).

If the lobbying role played by consultants was controversial, the work of those employed by government raised other issues. There were revelations of conflicts of interest at regular intervals in the project. This is a common feature of privatisation processes (Hodge and Bowman 2006, 118-120) and it provided an ongoing subplot to the bidding process. Consulting firms such as those employed by Finance, due to their size, have at any one time, a range of clients. As a consultant explained, this meant that there were "familiar faces on the other side of the (negotiating) table" (Interview: 9). The example of Pröpper, the SPD and RWE consultant has already been mentioned. Just as damaging was the revelation in March 1999 that Merrill Lynch was also working for Enron on the planned flotation of its water company subsidiary Azurix. This was the company through which Enron was formally engaged in the bidding process for BWB. It was observed that if Enron won the bid, it would increase the value of Azurix (Wiskow 1999). This brought criticism from, among others, the CDU (Berliner Mogenpost 1999b). Nevertheless, Fugmann-Heesing defended Merrill Lynch and they retained their role as managers of the process (Interview: 9).

There were further examples. Hengeler Mueller, the legal firm working for Finance, RWE, partners in the ultimately successful bidding consortium, were also one of their biggest clients. RWE also employed a consultant, Herbert Märtin, to lobby the SPD. He had been involved in a corruption scandal surrounding plans to develop an international airport in Berlin (*Großflughafenaffäre*) earlier in the 1990s. At the same time, Vivendi employed former airport manager Manfred Hölz to do the same job (Schulte 1999d). In July it was revealed that Credit Suisse First Boston, who worked for the RWE/ Vivendi/ Allianz bid, had at the same been working for the *Berliner Flughafen Holding* ("Berlin Airport Holding") of which the Mayor, Diepgen, was CEO (Spannbauer 1999).

Later in the process, there was yet another example of a potential conflict of interest. The President of the Constitutional Court, Klaus Finkelnburg, had to declare himself prejudiced when he was hearing the challenge to the Partial Privatization Law in the summer of 1999. This was because his firm had recently merged with White & Case, which had been working for the Economy Senate on the Partial Privatisation Law. The issue of conflicts of interest added a hint of scandal, common to privatisation processes, and probably reinforced the view that privatisation can be a "legalised form of corruption" (Hodge and Bowman 2006, 120). Certainly, it is clear that the governance structures which emerged through this privatisation process become increasingly populated by private sector experts and lobbyists. Consultants and large private firms were particularly prominent and became enmeshed in a complex web of relationships in which business and politics overlapped. Within these overlaps, interests were so entangled and activities so secretive that it is difficult to clearly determine who was working for the 'public' interest and who was working for private interests.

Resistance to the management of the process

A further problem for the management team at Finance was that other actors resisted their control of the process. The Parliamentarians – especially from the SPD (Schulte 1999c) – were concerned about the effects privatisation would have on water prices. The Unions and BWB Workers were still threatening to strike against privatisation even if privately they may have been looking for a deal regarding job security. With the CDU openly calling for the privatisation process to be halted, a group of Housing Associations threatened to challenge the privatisation on the grounds that it would lead to a huge increase in water prices (Riecker 1999c). These concerns are inherent to privatisation and the management team were unable to quell them: their attempts under the guidance of Fugmann-Heesing to define the process in terms of the sale price and manage the process accordingly were not entirely successful. As Saint-Martin (2000, 206) states, the democratic political process imposes limits on the influence of consultants, no matter how vital their expertise is.

Around the same time in March 1999, Parliament delayed a vote on the Holding Model, due to unease, particularly amongst the SPD members, over job security and water prices. Despite the fact that his party had voted in favour of the partial privatisation the previous year, the SPD Berlin Party Vice-Chairmen, launched an attack on Fugmann-Heesing's fiscal policy, questioning the sale of the "jewel" amongst Berlin's assets (Scharf 1999). This attack coincided with the SPD Parliamentary Party demanding that the Partial Privatisation Law be amended to include a provision that prices would not be raised until 2004 (Schulte 1999c).

This period seems to have been the most threatening to the privatisation process. It can also be interpreted as a set of resistances to the depoliticisation of the process. As a result of these resistances, two major concessions emerge. First, the Coalition Government came to an agreement on water prices – fixed until the end of 2003 (extended beyond the original CDU proposal of 2001) (Berliner

Zeitung 1999c). Second, the Berlin government signed another, more formal agreement with the BWB workers on the 13th April 1999, guaranteeing that no jobs would be lost for 15 years.

The way in which these agreements were made reveals much about the way in which the management team attempted to run the process. Although this set of actors was often successful in detaching itself (and their negotiations with the bidding companies) from the broader political system, they did not break free of its constraints. The process remained political and there were successful interventions by other actors. Of course, resistance and reform were usually within the terms of the privatisation. Nonetheless, these concessions, however cynical, short-term and illusory they may have been, reveal a political process beyond the control of the management team. Their attempts to define the privatisation process as far as possible in financial terms (to attain a high sale price) were, at certain times, successfully disrupted by other actors in the political system. The opposition from the CDU leadership and SPD Parliamentarians can be interpreted as being, in part, a contestation of the new quasi-institutional arrangements through which the privatisation process was being managed.

Finance- Parliament

If it was hard to access information within Finance and the government it was even more difficult for those outside. The SPD parliamentarian stated that it was "difficult to get the right information" from Finance and he saw the consultants as instrumental in this (Interview: 16). Like Fugmann-Heesing, the consultants were "finance-oriented" and the "sale price was everything" (Interview: 16). The parliamentarian also noted that both parties employed a large amount of consultants in the debate over privatisation models but that those at Finance were the "inner circle" (Interview 16).

According to this interviewee, in the government's dealings with the Parliament it was the consultants, and not members of the government who were "selling privatisation to Parliamentarians" (Interview: 16). i.e. making presentations, explaining aspects of the deal. "Educating the client" is how one of the consultants put it (Interview: 9). This is a clear example, of the extent to which the consultants took on – through their expertise in this field – functions normally associated with politicians or civil servants. Overall, the SPD parliamentarian argued that consultants were "enriching the power of the Finance Senate": "they were the experts and we were just small politicians" (Interview: 16).

There were those within both parties opposed to privatisation (though mainly within the SPD) and many others who opposed the model adopted to achieve it. As a result, there were tensions between the consultants and the

Parliamentarians but that they were "tactical enough not to get involved in political conflicts" (Interview: 16). The consultants were critical of the Parliamentarians in interview. During these meetings the politicians asked only one question: "what will happen to water prices?"(Interview: 9) Beyond this, he stated, they "didn't seem to be interested in the details" (Interview: 9); talks were very "one dimensional" (Interview: 11). It may have been true that Parliament was not rigorous in its scrutiny of the details of privatisation but, given the limited access to information, it could be argued that this was a 'managed' outcome. Furthermore, such was the complexity of the privatisation it is arguable that, apart from the consultants, few could truly understand the Holding Model and other aspects of the law (Beveridge and Hüesker 2007).

Even for those with experience of water privatisations, the BWB partial privatisation was challenging. Senior representatives of Thyssen, Suez, RWE - Umwelt and Severn Trent all believed that it was the most complicated privatisation model in the World (Schulte 1999a). The consultant who worked for RWE/Vivendi/Allianz on their bid, said that the final arrangements were a "total mess", one that "only legal experts can easily understand" (Interview: 8). Aspects of the privatisation deal are discussed below. At present it is worth adding that many others interviewed shared this view (Interviews: 2, 6, 7, 10)

Obviously the lack of scrutiny is not to be applauded, but their anxiety about prices for customers did lead, in the view of a consultant at Finance, directly to the fixing of prices for 4 years (interview: Etz). Although cynical in its short-term nature, with one eye on the forthcoming elections in October 1999, the fixing of prices beyond the next Parliamentary elections does nevertheless indicate that the Finance team could not simply impose their will on the privatisation process. With this agreement added, the Partial Privatisation Law was finally approved by Parliament on the 29th April 1999 (106- for; 67- against).

CDU, Unions and BWB Workers

If pressure from the SPD can be seen as a major factor in the fixing of prices for 4 years, then the CDU leadership can be seen to have engineered the deal on jobs with the Unions and Workers. Just prior to the signing of the contract, Landowsky publicly supported the Union's threat to strike if their positions were not guaranteed (Riecker 1999e). As stated Landowsky, and the CDU more generally, enjoyed close relations with the Unions and BWB workers. The negotiations over a compromise with the Unions and Workers had been ongoing from 1997 and the Senate's formal decision to privatise BWB on the 8th July 1998 in the Holding Model came the day after the conclusion of the first *Vertrauen Vertrag* ("Contract of Trust') between the workers and the Berlin government. A similar, more formal, deal was signed on the

13th April 1999 despite Fugmann-Heesing's opposition. In fact, a journalist for Die Tageszeitung (Bollmann 1999) claimed that the deal was signed by the CDU Senator for Interior (officially responsible for public sector employees) without Fugmann-Heesing's agreement. As in 1993, when Landowsky had sealed a compromise between the Berlin government and the Unions to ensure the commercialisation of BWB and other public companies, the CDU had accommodated workers interests within policy reforms (Passadakis 2006, 10). Perhaps strangely then, it was the CDU leadership that displayed a tendency towards a corporatist form of decision-making, while the SPD leadership, and Fugmann-Heesing in particular, resisted it. In fact, this style of policy-making, bargaining with the Unions, was one of the targets of Fugmann-Heesing's reforms (Interview: 14).

The Unions and BWB Workers were, however, permitted only a limited role in the bidding process. Again, like the Parliamentarians, they had little access to information or direct contact with Finance. The official contact they had with Finance was generally through meetings with the consultants about the Holding Model (Interview: 17). Despite these meetings, the Union representative stated that most of their information came informally from the Parliament – in particular the CDU Parliamentarians (Interview: 17). The Unions were permitted just one official meeting with every bidding consortium.

This mirrored the involvement of the Management Board. When asked about the treatment of the Management Board, one consultant confirmed their marginalisation when interviewed and was entirely unrepentant, arguing that they had "very little credibility", that they had badly managed BWB and, as "political appointees", they lacked legitimacy (Interview: 11). The consultant thought they "had a curious understanding of their own role in the process" (Interview: 11). To him, and the management team, they had no significant role to play in the development of a new BWB. Their attempts to intervene were seen as impediments to the achievement of privatisation. They were, the consultant stated, "total idiots", "very disruptive all the way" and "openly boycotted the process" (Interview: 11).

The negotiations with the private sector companies

Although the broader political process impinged on their attempts to manage the privatisation, Finance's control of the actual negotiations with the bidding companies seems to have been effective. Indeed, even in March, with a range of threats to the privatisation apparent, Finance emphasised that the confidential negotiations would go on and, furthermore, that such threats would not influence them. The press were informed that contracts agreed with investors would be signed without Parliamentary approval and that the Parliament had nothing to do with the negotiations with the bidding companies (Voss 1999).

This section examines the bidding process, bringing in the accounts of participants in the successful RWE/Vivendi/Allianz bid. Having established that negotiations were largely cordoned-off from the rest of the political process and that the priority of the Finance Senate was the sale price, the aim is to reveal the extent to which crucial aspects of the final arrangements were determined in these negotiations. This shall then be compared to the assertion made in the ready-made 2 account that all key aspects had been agreed by the government prior to their involvement.

First, it is necessary to emphasise the extent to which the bidding process was – from the Berlin government's side – managed by the consultants. As stated, they were officially the only contact between the public and private sides during negotiations. Throughout the entire bidding process the consultants were ever-present, be that in the 'Data Room' or in direct negotiations with potential buyers (Interview: 9). The consultants would then report directly back to Fugmann-Heesing on how things had proceeded. Importantly, they only reported to Fugmann-Heesing and no one else either in Finance or the government (Interview: 9). They were the means through which Berlin's interests were represented and were always trying to "push up the price and reduce criticisms from the bidding companies" (Interview: 9). Fugmann-Heesing and, on at least some issues, other members of the Executive may have had the final word, but much of the information on which they based their decisions came from the interpretations of the consultants.

The consultants from Finance were up against, not merely the representatives of the companies, but the huge teams of consultants they employed on the bid. For example, the RWE/Vivendi/Allianz consortium employed 80 consultants at the high point of the bidding process (Interviews: 13, 8). These were complex, often tough negotiations (Interview: 9). They were also incredibly expensive for both sides due to the amount of consultants employed and the length of time they took. This raised the stakes even further – as both the Berlin government and the bidding companies invested great amounts of money in simply conducting the bidding process.

In total, a consultant for the RWE/Vivendi/Allianz consortium thought they spent $16.7 Million (DM30 million) (Interview: 8). This was a reflection of both the scale of the task (the complexity of privatisation) and, of course, the resources the consortium was willing and able to devote to "*the* privatisation worldwide" of the period. Specific help was required for German Law, Financial Modelling, PR and Environmental Management (Interview: 8). The key consultants employed by RWE/Vivendi/Allianz came from some of the biggest consultancy firms in the world. KPMG, who dealt with tax and tariffs, are one of the so-called

'Big Four' in global consultancy industry (Hodge and Bowman 2006, 101). Other names were also high profile: Credit Suisse (Banking) Freshfields (Law) and H.P.C (Environment). With these household names of the business world sitting across the negotiating table from each other, the degree to which the BWB privatisation process came to resemble a high profile business sale is striking. This was Fugmann-Heesing's aim, the consultants' job, and it seems it was achieved.

Nonetheless, it was not simply a business deal but a privatisation and therefore potentially more controversial. A lot of time was devoted to lobbying the media, politicians and Unions (Interview: 8). The Vivendi executive stated that they also employed a media agency as it was essential to "let them know who we are, what we are, what we do" (Interview: 13). The consultant working for the consortium stressed the importance of Jean Marie Messier, the Chairmen of Vivendi (Interview: 8). during this period, he was diversifying the company into media sectors, using the steady and secure income from utilities to fund buy-outs. The consultant argued that he needed a big project to prove himself, and "Berlin was his baby" (Interview: 8). This was important as he "put himself and his best people from France into the project" (Interview: 8).

Messier seems to have spent a lot of time talking to politicians in Berlin and, in particular, addressed their dreams of making Berlin a global city. As would be shown in the final contract, part of this dream for Berlin was to make it a centre for IT, new media and entertainment. This was a happy convergence of interests between Berlin's political elite and Vivendi. However, its importance should not be over emphasised. A consultant at Finance stated that Fugmann-Heesing had no interest in the media side of the deal (Interview: 11), though it was conceded that the Mayor did. By this time Vivendi owned Babelsburg studios to the west of Berlin and a number of prominent politicians were invited around as part of the charm offensive (Interview: 8). They were, apparently, "seduced by the media glamour" of the event (Interview: 8).

Was all this important? The consultant working for the winning consortium thought it was as such work creates an atmosphere in which success is more likely (Interview: 8). This was part of their strategy for "creating a good feeling" around the Vivendi/RWE/Allianz bid (Interview: 8). They believed it was vital to create a sense that Vivendi/RWE/Allianz consortium was "confident, competent and professional" (Interview: 8).

A trade-off between sale price and long-term guarantee on profits

Regardless of the PR campaign which accompanied the bidding process, the financial aspects of the deal were the focus of discussions and these were conducted in secret. The Vivendi executive confirmed the main focus of discussions was the sale price and tariffs (Interview: 13). The environmental and technical issues of the deal were secondary: it was "finance-driven" and they had little contact with the environmental and technical people within the Berlin government and BWB (Interview: 13).

Finance wanted the money quickly but what did they have to negotiate with? As stated in the previous chapter, the bidders were willing to pay a strategic price for BWB because of its perceived global potential. They were, then, willing to take on the risks of some of BWB's loss-making subsidiary companies. They were also willing to accept the fixing of prices for a limited period.[31] But there needed to be some means of ensuring a return on their investments. One line of negotiations centred on the form of privatisation – the Holding Model itself, with the bidding companies making "side-offers" (proposing alternative models) right up until the final contracts were signed (Interview: 13). One early offer from the RWE/ Veolia/ Allianz team included separating ownership from assets (Interview: 13). Also, at a later stage, RWE/ Veolia/ Allianz offered to lease BWB from the government and they even offered a higher price than the final bid accepted, in a deal which included selling back elements of BWB to the Berlin government (Interview: 13).

The Vivendi executive stated that the government were pretty unwavering in their desire for the Holding Model (Interview: 13). It was not, however, Fugmann-Heesing's first choice. According to the civil servant at Finance she had wanted full privatisation from the beginning (Interview: 14). The Holding Model had, however, been approved by the government and was central to the Partial Privatisation Law passed by Parliament on the 29th April 1999. It was the outcome of a relatively long process of negotiation and was the compromise necessary to achieve support for privatisation. In this sense, the management team were restricted in their negotiations: they did not have the freedom to change the overall shape of the privatisation. Crucially, however, they were able to negotiate with the bidding companies on vital details of the privatisation within the drafting of the Partial Privatisation Law and, then, in the negotiations over the public and confidential contracts between the buyers and the State of Berlin.

'R+2'

Negotiations can be said to have hinged on the sale price (and promised investments in BWB and Berlin) and the tariff calculations. As the CDU politician very openly stated, the key aim of a high sale price, to service the city's debt (and interest paid on the debt), could only be achieved by guaranteeing the future income

---

[31] Fixing prices for short period is not that uncommon in WSS privatisations *e.g.* Sofia, Bulgaria (Lobina 2001, 5).

of the private partners (Interview: 15). This is where the discussion of the 'R+2' Formula became vital in the drafting of the Partial Privatisation Law. Later in the contract negotiations, the length of the contracts and mechanisms for arranging the management structures between Land Berlin and the private partners within the Holding Model became important.

These details are part of the confidential contracts but interviews with both the consultants for Finance and members of the RWE/Vivendi/Allianz bidding team reveal that these discussions defined far more than suggested in the ready-made politics 2 account. The civil servant was initially very clear that by the time the consultants were working for the Senate (from July 1998 onwards) almost all of the Partial Privatisation Law was in place. This version fails to reveal that crucial details of the privatisation deal were defined during the bidding period up until the signing of the final contracts (October 1998-October 1999): in negotiations between the bidding companies and Finance, with Economy on the margins. In effect, consultants can be seen to have played a key role in devising the Partial Privatisation Law before it was sent to Parliament on January 5th 1999 (where it was eventually accepted with only minor alterations). Furthermore, their leading role in the negotiations of the contracts between Land Berlin and the private partners (October 1998-September 1999) placed them at the centre of decision-making process of a political and legal character. Both of these decision-making processes were secret, not subject to parliamentary scrutiny nor the rules and norms of the formal political system.

From the accounts given by the consultants leading the bidding process, it seems clear they had a direct role in shaping the Partial Privatisation Law. One stated in interview that they had two main official responsibilities: first, preparing the sales documents – describing and valuing the company; second, designing the framework for privatisation – assisting with the drafting of the Partial Privatisation Law (Interview 11) according to the Holding Model (which had been decided upon by the time Fugmann-Heesing assembled the team of consultants).

One of the consultants stated that drafting the Partial Privatisation Law was complex given that wastewater is legally a municipal obligation in Germany and must therefore remain in public hands (Interview: 11). Despite the challenge this provided in terms of drafting the law, it had one great advantage for both public and private partners. Unlike freshwater services, no VAT would have to be paid on wastewater. If BWB had been turned into a Plc., for example, all VAT would have to have been paid and this would have resulted in higher prices on wastewater services (Interview: 11). This was one of the key reasons why the Holding Model was chosen (Interview: 11). The main functions of the company continued to be the overall responsibility of the public partners, while the day-to-day management was undertaken by the private sector (Interview: 11).

Another challenge for the consultants was to construct a tariff system. There was no precedent in Germany for establishing a public-private tariff system for water and wastewater services. Indeed, it was stated publicly at the end of February 1999 – almost 5 months into the bidding process – that water tariffs were still being negotiated (Zeiger 1999). The fundamental question facing them during the bidding process was: how much profit should the private (as well as the public) partners be allowed to make? This is where the consultants played the dominant role (Interviews: 9, 11). One aim of the tariff system was to determine the profits of the owners. This was a critical part of the Law because the public-private company would enjoy a monopoly within the boundaries of Berlin. After much consultation with both politicians and the bidding companies the consultants came up with the controversial 'R+2%' formula, which was written into the Partial Privatisation Law.[32] To re-cap, 'R+2%' refers to 'revenue plus 2%'. 'R' is calculated in relation to the average percentage revenue made from 10 year, low-risk Stock Market Bonds over the previous 20 years. Once this percentage had been calculated, for example for 2004 it was 6% (Werle 2004), then 'R' would equal 6% of the revenue of BWB. 2% would then be added on to determine the profit rate for both the public and private owners.

The Veolia executive, the civil servant at Finance and the senior SPD politician all denied in interviews that the formula guarantees the private partners profit (Interviews: 12, 13, 14). In contrast, one of the consultants working at Finance very openly stated that it was designed to ensure around 8% profit (Interview: 11). Given the controversy the 'R+2' Formula has provoked in Berlin it is worth stressing what was said in interview: "it effectively guarantees profit" (Interview: 11). 'R+2' does not guarantee a specific amount of profit (as the calculations for 'R' may change every year) but it provides a guide as to how much that profit should be. The consultant said that it was hard to come up with an "acceptable figure" for this profit and they spent much time examining other private water companies and other utility sectors around the world (Interview: 11). The figure of 8% mirrored profits generally made by water companies in the UK at the time (Interview: 11). It provided an acceptable comparison for the bidding companies.

This effective guarantee was particularly important to the private sector companies given the risks they were taking on with BWB's subsidiary companies (Interview: 4). The confidential contracts agreed alongside the Partial Privatisation Law make it near-impossible for either the public or the private partners to withdraw. In a sense the sale price can be seen as a loan, with the terms of re-payment set in the guarantee of annual return over the 29 years of the contract

[32] R+2 was only one part of the water tariff system. It is highlighted here due to the controversy it has provoked. For a broader discussion of the tariff system see Hueesker 2011.

(with the state of Berlin also, in theory, making a profit). As will be discussed shortly in more detail, 'R+2' was successfully challenged by the Opposition Parties in the Constitutional Court. In October 1999, the Court ruled that the means of calculating 'R' were acceptable – it was fairly common to use this means of calculation for municipal companies in Germany. However, it saw no grounds for a public-private monopoly to be able to simply add 2% on top. More generally, 'R+2' guarantees profit without risk for the owners: the rate of profit is based much more on a calculation than on performance. Put simply, 'R+2' removed much of the risk in investing in BWB.

Returning to the ready-made 2 account

In his first account of the process, the civil servant at Finance stated that the Partial Privatisation Law had essentially been agreed before the consultants were employed (July 1998 – October 1999). The accounts provided by the consultants make it clear that crucial aspects of the Partial Privatisation Law were in fact being negotiated from October 1998 *onwards*. Indeed, key features of the law, the guarantee of a return on investments and the tariff structure, were in fact under discussion until around March 1999, a period of around 6 months. They were not finalised until the end of the indicative bidding stage so that bidders could have a clear idea of what they were buying into when they made their offers. Beyond these two important elements of the Partial Privatisation Law, the influence of the consultants is unclear. Nevertheless, there can be no doubt that they were crucial to the process of translating the Holding Model into the Partial Privatisation Law.

On being presented with the consultants' accounts of the decision-making process, the civil servant backtracked in interview and accepted that they were involved in designing the 'R+2' Formula. However, in contrast to the consultant's version above, he stopped short of giving them full responsibility (Interview: 14). Nevertheless, with this step, the line drawn in the ready-made accounts between 'political' decision-making and the commercial bidding process blurs. The Partial Privatisation Law was described by a consultant working for the successful consortium, as a compromise between political parties conjured up by the consultants involved (Interview: 8). Additionally it appears they took control of most aspects of the day-to-day running of the bidding process (1998/1999) and effectively devised the strategy the Finance Senator employed to deal with not merely the bidding companies, but the media, the Parliament, the rest of the government, the BWB Management Board and the Unions. Did they, then, become political actors? One of the consultants admitted that they "became a component in the political process" (Interview: 9), while the other thought that they did become political actors, but "only by default" (Interview: 9).

RWE/ Vivendi/ Allianz win the bidding process

By March 1999, there was a second short-list, with only the genuine contenders left: RWE/Vivendi, Suez/Thyssen (with Mannesmann Arcor replacing the latter in mid-March) and the Enron subsidiary Azurix (and HeLaBa: Hessische Landesbank). According to Zimmermann, the real contest was between the two consortia containing the French companies, with Enron there to encourage further competition (interview: Zimmermann). By the middle of May, Suez and Mannesmann Arcor had been joined by Bankgesellschaft Berlin (Die Tagespiegel 1999b). It is unclear why exactly they joined, though for Suez having a Berlin-based partner may have been seemed strategically good and it is likely that Bankgesellschaft had offered to help finance a bid.

Although the sale price was the main priority of the Berlin government, it can be assumed that by the time of the second short list, the indicative offers of those left were all of an acceptable level. At this stage, the "whole package" (Interview: 9) was becoming more important to negotiations; the so-called "soft factors" designed to make a bid more attractive (interview: 8). For example, the RWE/Vivendi/Allianz consortium developed slogans such as "Not just saving jobs, but creating new jobs" and "We will create 1000 new jobs for young people" (interview:8) to appeal to the Berlin government. As a consultant on the RWE/ Vivendi/ Allianz project stated such clear messages are important in winning over politicians as they can use them when presenting the privatisation to the public (Interview: 8).

In these so-called soft factors the terms of the global city discourse are also apparent. The consultant for RWE/ Vivendi/ Allianz stressed that at end of the 1990s there was huge hype surrounding the media, new media and the internet and Berlin wanted to be part of it (Interview: 8). As part of their offer for BWB, RWE/Vivendi/ Allianz promised to help construct the Media-Tech to house a German film and TV archive in the Sony Centre at Potsdamer Platz (Interview: 8). In total they offered $4.8 Million (8 Million DM) to Berlin to get this started. Is there a danger of over-emphasising its importance? Zimmermann the consultant working for RWE/ Vivendi/ Allianz expressed his "surprised" by the length of time spent discussing items such as Vivendi's promise to invest in the Media-Tech (Interview: 8). To further highlight the importance of soft factors, the consultant also alleged that Enron made an agreement with Delta Airlines that they would provide direct flights from Berlin to the USA if Enron won the bid (interview: 8). This played to the ambitions of the politicians desire to connect Berlin to the key centres of the global economy.

Interviewees were divided as to the importance of such offers towards the end of the bidding process. The consultant for RWE/ Vivendi/ Allianz stated that the non-water agreements in the contracts were "not just gimmicks" (Interview: 8).

In contrast, one of the consultants at Finance stressed that the non-monetary non-water issues were "just for the gallery", to ensure it became "a popular decision" (Interview: 11). For him the non-water and non-monetary aspects of the deal were "overall not that important" and he was quite dismissive of the claim that they were surprisingly focused on the media side of things: "Fugmann-Heesing didn't give a damn" (Interview: 11). The consultant did, however, concede that Mayor Diepgen was interested in this aspect of the negotiations (Interview: 11).

However, when questioned about the assessment of bids, the other consultant working for the Finance Senate made the general point that, in privatisations of WSS services, non-monetary factors can often outweigh the monetary factors (Interview: 9). This is in part because privatisation has to be presented to a public who are normally unconvinced of its benefits (Interview: 9). Furthermore, politicians always have other interests. In the BWB case, this consultant stated that once they came to the final stage of the bidding process, the criteria presented to the three remaining companies included "what else you will do for Berlin" (Interview: 9). According to the civil servant at Finance they asked three key questions of the bidders (Interview: 14):

1. How will you expand BWB's international operations: how will you make BWB a high profile brand?
2. How will you make the – failing – 'Berlicom' (the telecommunications subsidiary of BWB) a success?
3. More generally, what is your strategy for running BWB – what assurances can you give us?

The civil servant stated that on all questions RWE/ Vivendi/ Allianz were stronger and their offer was the highest (Interview: 14). As a result, on the 14 June 1999 RWE/ Vivendi/ Allianz won the bidding process with a bid of $1.96 Billion (DM3.3 Billion) and signed a range of contracts with Land Berlin.

Within weeks of this announcement, however, there were rumours that Suez had actually offered $110 Million (DM200 Million) more than the RWE/ Veolia/ Allianz. Though they never actually admitted it, public statements made by Mayor Diepgen and SPD leader Böger did not directly deny the allegation. Diepgen stressed that the whole finance package was just as important as the overall price (Schulte 1999e). Böger stated that the price was "appropriate" and the total package played an "essential role" (Bruns 1999).

Is it possible that – right at the end of the process – the objectives of Finance and Fugmann-Heesing changed or were compromised by non-monetary issues? One of the consultants stated that the RWE/ Vivendi/ Allianz consortium was the highest bidder and the deal was "only about the price and the price was good"

(Interview: 11). Certainly the sale price of $1.96 Billion has been generally seen as high (Lanz and Eitner 2005a, 4). Still, the Vivendi executive stated that "it was possible" that the Suez consortium offered more though this may have been an attempt to allow the rumour to persist (as it suits the private partners of BWB to claim that their all-round package for Berlin was better) (Interview: 13). The civil servant at Finance was, however, very detailed in his denial. The rumour stemmed from Suez, furious at losing out to Vivendi (interview: 14). In interview the civil servant provided details of the final three bids:

1. Enron-led consortium: c. $1 Billion (DM 1.9 Billion).
2. Suez-led consortium: c. $1.17 Billion (DM 2.2 Billion) (including a sale and lease back system).
3. RWE/ Veolia/ Allianz: $1.96 Billion (DM 3.3 Billion).

(Interview: 14)

Apparently, there was, then, a difference of more than DM1 Billion between the bids of Suez and RWE/ Veolia/ Allianz. Perhaps the gap between the two bids can be attributed to the financial strength of Allianz and the lack of a comparable financial institution in the Suez-led consortium. It could also be explained more simply by Vivendi's greater willingness to pay a strategic price for BWB. Two years later Vivendi were criticised for vastly outbidding their rivals for a WSS concession in Prague (Lobina 2001, 6). Like Berlin, they saw Prague as an ideal base for expansion in Central and Eastern Europe (Lobina 2001, 6). The civil servant at Finance did say in interview that Suez was desperate for more time to raise investment for a higher bid (Interview: 14). Furthermore, while Enron were calm about the rejection, Suez threatened the Berlin government with legal action as soon as they realised the process was going to be closed (Interview: 14). They sent a letter to the Berlin government, claiming that their bid was higher, and their proposal better and that the decision was unjust (Interview: 14).

Certainly it makes little sense that the Berlin government, given its financial problems and the extent to which the management team had attempted to ensure a high sale price throughout would – at the last minute – choose the second highest offer. Still, this doubt underlines the extent to which the process was secretive and controversial. It could be argued that had the bidding process been less secretive, with, for example, all final offers being made public, it may have ended on a less controversial note. However, the bidding process had been managed as a 'commercial' process, one run according to the norms of the business world. For this reason all negotiations between the management team (on behalf of the Berlin government) and the bidding consortia remained confidential. The sale price of the publicly owned water company, the specific agreements regarding investment in the loss-making Schwarzepumpe SVZ ($106 Million/ DM 200 million) and investments in Berlin (*e.g.* $4.25 Million/ DM8 million for 'Media Tech' project) were decided apart from the rest of the political system. Pressure from actors within the wider political system had, however, shaped aspects of the final deal as the list below shows: the 'guarantee' of no job losses for 15 years, the fixing of prices for 4 years and the $164 Million (DM310 Million) to be directly invested in the Berlin economy.

Revising the Partial Privatisation Law: the legal challenge and the confidential privatisation contracts (June – October 1999)

These contractual agreements supplemented the Partial Privatisation Law. June 14th 1999 did not, however, mark the end of the process. Two parallel processes were apparent: one public, conducted through the formal political system and the other, secret, conducted between the management team and the RWE/ Vivendi/ Allianz consortium. At no stage in the privatisation process was the cleavage between the formal and extra-formal processes of policy-making so apparent. The rule-based formal process of the political system continued to shape aspects of policy but key decisions were made elsewhere. This occurred even beyond the institutions of the executive: in the informal, ad hoc arrangements centred on the Finance Team.

At this time, the Partial Privatisation Law was being challenged in the Constitutional Court. Shortly after the Parliament had approved the Partial Privatisation Law with minor amendments on the 29th April, the opposition parties (PDS and the Greens) launched a legal appeal in May. Given the Parliamentary majority enjoyed by the CDU-SPD coalition, this was one of the few genuine opportunities for resistance provided by the formal political process. Their challenge rested largely on the claim that the 'R+2' formula was unconstitutional: that a public or partly public enterprise should not be legally bound to provide profit returns to its owners.

Although a deal had been agreed, negotiations between the Finance Team and RWE/ Vivendi/ Allianz continued. These secret discussions dealt with defining the more exact details of how the partially privatised BWB would function in terms of daily management and the selection of the Management Board. They not only supplemented the Partial Privatisation Law but actually reformed aspects of it. Of perhaps even more significance, these additional contractual agreements pre-empted the Constitutional Court's ruling. In other words, the management team at Finance continued to work as they had done before. Despite the signing of contracts and public announcements – confidential negotiations with the bidders persisted until the formal partial privatisation of BWB on the 29th October 1999.

Table 6.4: Key privatisation agreements (made public):

1.	Prices fixed for 4 years: 1999 to 2003.
2.	No job losses for 15 years.
3.	Economic independence of BWB ensured i.e. it must remain a separate entity from RWE and Vivendi.
4.	DM 5 billion investment in the first 10 years and penalised if targets are not met.
5.	Stated aim to make BWB international player in water markets.
6.	Berlin as an international centre of competence for water management.
7.	Developing new jobs.
8.	DM 310 Million to be directly invested in the Berlin economy: establishment of a 'future fund' for technology development
9.	If put on stock exchange (a possibility at the time) 10% of shares would go to the consumers, and 10% of shares to employees.
10.	Sale price: $1.96 Billion (DM 3.3 Billion)
11.	$106 Million (DM 200 Million) to deal with Schwarze Pumpe SVZ
12.	DM 3.1 billion to Land Berlin
13.	Investment in 'Media-Tech' ($4.25 Million/ DM 8 million).

Source: translated and adapted from a press statement, Berlin Senate 18/06/1999.

Alongside the main secret *Konsortialvertrag* (Consortium contract: 14th June 1999) there were three additional contracts:

1. *Kauf- und Uebertragungsvertrag* ("Sale and Transfer contract").
2. *Vertrag über eine Stille Gesellschaft zwischen Beteiligungsgesellschaft und Holding AG* ("Contract re: One Silent Partnership between the investment company and the Holding").

3. *Vertrag über zwei Stille Gesellschaften und zur Begründung einer einheitlichen Leitung zwischen Anstalt und Holding AG* ("Contract re: two silent partnerships and for the establishment of a uniform control of the public law institution and the Holding AG" (PLC)

Aware of the legal challenge, negotiations between the Finance management team and RWE/Vivendi/Allianz resulted in a fundamental amendment to the main consortium contracts. In the *Konsortialvertrag* contract, the Berlin government ensured that, regardless of the Constitutional Court's decision on 'R+2', the private partners would receive the same profit rates outlined in the formula. This contractual agreement did, in effect, pre-empt due legal process. When the Court eventually ruled that 'R+2' was unconstitutional on the 21st October 1999, it was removed from the Law.[33] In effect, profits for the public and private owners are determined by the secret agreement in the *Konsortialvertrag* contract. The Berlin government had, without the knowledge of Parliament, contractually obliged itself to an agreement on profit returns even if it was judged to be unconstitutional.

The other contracts also raise concerns. The Partial Privatisation Law outlines the overall framework for the partial privatisation of BWB according to the Holding Model. However complicated, this model was designed to balance the public and private partners' powers in decision-making slightly in favour of the public partners. Retaining overall public control, whilst ceding the day-to-day management to the private partners, was crucial to gaining the agreement of the SPD. The additional contracts, however, have a much broader scope. They deal not only with day-to-day management but with the control of the BWB management structures. It is also important to stress that the contracts are unlimited with the possibility for review and termination only in 2028 (29 years after the privatisation). As they are confidential they cannot be accessed by Parliamentarians let alone the public. In effect, the contracts legally embed a management and policy-making structure detached from parliamentary scrutiny for a period of 29 years (Beveridge and Hüesker 2007, 22).

The Holding Model entailed the foundation of a *Beteiligungsgesellschaft* (the Investment Company set up by the private partners) in June 1999, which took over 49.9% of the shares of the BWB Holding Plc. (majority-owned by the State of Berlin in partnership with the private partners) which had been formally established in July 1998. These new companies, *Beteiligungsgesellschaft* and Holding, then entered into three silent partnership agreements giving them a total of 49.9% control of the continuing public-law entity BWB (*Anstalt öffentlichen Rechts*). It was for this share of the *Anstalt*, that the private investors – through the *Beteiligungsgesellschaft* and the

[33] The court also ruled that a complex efficiency incentive and a particular section on decision-making between the *Anstalt* and the Holding were unconstitutional: see Hueesker 2011. One consultant believed that the former was "killed on political grounds" (Interview: 11)

Holding- paid the state of Berlin $1.8 Billion (DM3.3 Billion). The state of Berlin therefore kept a majority share of the Holding and the *Anstalt*. This was a key condition of the Holding Model. In return the private investors were permitted the lead in the day-to-day business of the partially privatised BWB.

The secret contracts altered the balance outlined in the Partial Privatisation Law. In particular they shifted control of nominations to the Management and Supervisory Boards of BWB to the private sector partners. Mirroring the means through which the privatisation was implemented, this was achieved through the establishment of informal decision-making institutions alongside those formally established in the Partial Privatisation Law. For example, the *Konsortialvertrag* (Consortium contract) legally established the *Vorstands Ausschuss* (Board Committee) which effectively revised part of the law.

The law stipulated that the public and private partners were allowed to propose two members each to the Executive Board of the *Anstalt* (responsible for the day-to-day management of the company). The *Konsortialvertrag* revised this by shifting responsibility for selection of the Chairman – away from the Berlin government – to the *Vorstands Ausschuss*. This institution consists of three members: one each from Veolia, RWE and the Berlin government. The government is thus outnumbered by the private partners. Beyond effectively giving the private partners control over the appointment of the Chairman, the Chairman's position within the Executive Board was strengthened in the *Konsortialvertrag*. The Chairman was given the deciding vote if no agreement was possible between the four members of the Executive Board. If the Partial Privatisation Law had aimed to balance public and private power just in favour of the former, then the *Konsortialvertrag* sought to shift the balance to the latter.

BWB post-privatisation

Shortly after the deal with RWE/ Vivendi/ Allianz had been agreed in June 1999, the Mayor was effusive in his praise. It was a "masterpiece" in social and economic terms (Reitemeier 1999), an "exemplary piece of future-oriented politics for Berlin" (Der Tagesspiegel 1999c). On formal completion of the privatisation on the 29th October 1999, the Senator for Economy stated that the government's objective was to make BWB "competitive" (Fischer 1999). More than ten years later, it is difficult to argue that these claims have been fulfilled. Rather, they have been revealed to be just what they are: partial claims about the merits of the deal and the advantages of private sector expertise. The former claim rests on assumptions about how the policy was made (how the Partial Privatisation Law and contracts were agreed), while the latter rests on neo-liberal assumptions that the private sector partners would improve the performance of the company, particularly within international water markets.

The price paid by the private investors was large ($1.8 Billion /DM 3.3 Billion), and this money was used to immediately pay-off some of the city's massive debts. The privatisation contracts do, however, guarantee profit each year regardless of the performance and actual profits made by the company. In years when the performance of BWB is worse than expected in the calculation period, which has often occurred as a result of the city's decreasing water consumption, Berlin must either raise the tariffs or meet the requirements of the guaranteed revenue by withdrawing money from the State budget. The State of Berlin – as the 50.1% – majority shareholder has also been forced to abstain from its own share of the profits. This requirement has already resulted in the transfer of millions of Euros from the state's budget (see Hüesker 2011).

Despite the job security agreement between the government and the BWB workers (which prohibits business-related lay-offs until 2014), the number of BWB employees has been reduced: from over 7000 in 1999 to around 5000 in 2006 (Passadakis 2006). It is also, for example, difficult to claim that privatisation has brought wider social benefits. Prices were fixed in the contracts until 2003 but between 2004 – 2008 prices rose by 25% (Hüesker 2011). This can be linked directly to the unconstitutional application of the 'R+2' formula. Using the revenue to calculate the profit encourages investment in BWB. For this reason investment in BWB (up until 2008) has been above that agreed in 1999 ($300 Million rather than $270 Million). However, because of declining consumption such investment is not actually required. In fact, it has increased problems of over-capacity (Hüesker 2010). As a result, the present Senator for Economy, Harald Wolf, of *Die Linke*, has been attempting to block investment in BWB (Interviews: 2, 7). By doing this, the aim is to prevent unnecessary investments but also to reduce price increases by keeping the revenue down and, in turn, the annual profit the private partner is entitled to receive.

In commercial terms, BWB is now less of a global player than it was under public management. In fact, BWB now focuses only on its core business of WSS management in Berlin – the international, multi-utility and services functions have been dropped (for discussions see Werle 2004, Passadakis 2006 and Hüesker 2011). In part this is because the first few years of commercial investments proved to be just as unsuccessful as those under public management. For example, the telecommunications company, Berlikomm, which was specifically earmarked for development, ran up debts, suffered job losses and was sold. Similarly, Schwarze Pumpe (SVZ), continued to be just as problematic under private management, with the sale of the company collapsing in 2001 and further losses being incurred by BWB. Eventually the company was sold in 2002 for €1 (for detailed discussions see Passadakis 2006 and Hüesker 2011). Most embarrassing of all, perhaps, was the liquidation of BWB's multi-utility business enterprise, Avida (providing telephone and electricity services), within weeks of its establishment in December 2001 (Werle 2004).

Thus the partially privatised company essentially performs the same functions BWB did prior to its move into global water markets in 1995. Drinking water quality and environmental standards have been maintained – they were already high under public management (Hüesker 2011). The key difference now is that private partners must be paid a profit every year regardless of performance. Furthermore, the contract duration is unlimited, with the first possibility for review and termination being 2028, nearly 30 years after the privatisation. Additionally, the government and the private partners have devised secret rules and procedures which now determine key aspects of water management.

Conclusion: assessing the accounts of the privatisation process

The ready-made 2 account of the process described the consultants as mere managers of the bidding process. This has been shown to be at best incomplete when analysed with reference to the timing of key decisions and accounts provided by other actors. Like the first ready-made account, it also rests on an image of how policy is made and realised, albeit a more credible one in which private sector actors participate in arrangements of governance. The first account sought legitimacy and accountability (if not transparency) in the model of the classical-modernist system, the second account alluded to notions of the managerialism in government: of the traditional processes of government adjusted to the 'need' for private sector expertise in governing contemporary societies.

The politics in the making account of the BWB partial privatisation has revealed a more problematic policy-making process. The management team at Finance were integral not merely to the 'commercial' aspects of the process but also to the political decision-making process. The image of decisions being made through the formal realm of politics was breeched: the major political decisions were not only made by institutions populated by elected politicians and civil servants bound by the duties of public office. Instead, a picture of governance in action has emerged, but one which is problematic in its mixing of the public and private and its departure from the official rules and institutions of the formal political system.

The emergence of this management team at Finance and their very calculated attempts to define the privatisation process in terms of a commercial deal and manage it accordingly were, at least partially, successful. In effect, a mainly enclosed decision-making process emerged alongside the formal institutional processes. The influence of this purpose-built form of governance and the roles of the consultants in particular, places a big question mark over the value and indeed the legitimacy of the formal institutions of the political system. This is not to suggest that these institutions were unimportant. In fact, the partial privatisation was still by-and-large conducted through them. Rather it is to suggest that the dynamics of the policy process were altered from the point at which the privatisation policy had been formally approved. A Steering Committee may have been established to coordinate the actions of government but the management team attempted to detach itself from it and the wider political system.

This dynamic was characterised as a series of duels between the management team and various actors and institutions within the policy process. There were clear examples of resistance to the Finance team, no matter how effective or genuine the resulting policy measures were. The Parliament's pressure to fix prices and the Unions' and BWB workers' threats to strike resulted in compromises that restricted the management team's room for manoeuvre and their attempted definition of the process purely in terms of the sale price. Other opposition, in the form of a legal challenge, also slowed down the privatisation, thereby preventing the quick sale that Finance wanted. Formal political powers (of the Parliament, the Coalition Government and the Constitutional Court) were, then, unmistakable. They did, however, have to be translated rather than merely applied – they were the contingent outcomes of the exchanges between actors, rather than the explanation for how those exchanges were conducted.

The management team was assembled to achieve a high sale price and achieve the wider objective of Fugmann-Heesing to bring global players to Berlin. The private sector consultants brought technical knowledge and experience of privatisation which could not be found in the public sector. They also had a way of doing things. They introduced practices from the commercial sector, notably the confidentiality agreement which effectively excluded the rest of the government and the Parliament from the negotiation of the sale price, the contracts and aspects of the Partial Privatisation Law. In doing so, the management team monopolised negotiations with the private companies. In many ways this suited the private sector companies. The management team at Finance may always have been pushing for a high sale price but in doing so they were less concerned than the Parliament and the CDU leadership about securing jobs and water prices. It is apparent from interviews and press reports that Finance saw these as impediments to the objective of a high sale price. Perhaps these deals were reflective of the Berlin "mentality", which the senior SPD politician wanted to change. Furthermore, the desire to avoid controversy (e.g. the confidentiality agreement and the strategy of restricting information flows within the political system) also suited companies such as Suez and Vivendi who had been embroiled in allegations of corruption in other privatisations.

To achieve its objectives, the management team, in its negotiations with the bidding companies re-wrote the rules of policy-making as it went along, even going as far as to pre-empt the Constitutional Court's ruling on the Partial Privatisation Law. The account paints a picture of a political system in which nothing could be taken for granted – no assumptions can be made about how policy is made and who is making it. As such the politics in the making account thus raises questions about the traditional institutions of the political system, of government conceived in managerialist terms and new forms of public-private governance. As the BWB case reveals, the presence of extra formal governance arrangements often undermined the very aims of the political system: to provide accountability and legitimacy.

Rather than referring to models of political decision-making, a more accurate summary of the implementation process would be that it reveals the emergence of yet another public-private policy-making elite so characteristic of policy-making in Berlin during this period. The consultant team at Finance was, of course, challenged by other actors but its composition reflected the merging of the public and private realms in new governance structures in Berlin during the 1990s. In this sense it was emblematic of the global city policy-making practices to which the last chapter referred: of the public-private Bankgesellschaft, the experts hired to help re-imagine and promote Berlin and the consultants advising the BWB Management Board. Seen within this context, the emergence of the management team was also part of translating the global city policy discourse. As was argued at the end of Chapter 5, the global city policy discourse provided not just a way of thinking about policy-making – a neo-liberal way of thinking – but it also provided a means of making policy – a neo-liberal means of practising politics. For these reasons, an assessment of the consultants, the Berlin political system and the BWB privatisation process should be placed within a broader examination of the effects of neo-liberalism on policy-making.

7. Assessing the BWB partial privatisation

Introduction

At the start of the study the partial privatisation of BWB and urban governance in Berlin were presented as being emblematic of broader political trends during the period. The BWB privatisation was characterised by the 'no alternative' study which accompanied many privatisations during this time, while the more general trends in governance in Berlin revealed the apparent 'necessities' of urban policy-making in the context of economic globalisation. It was argued that an analysis of both would reveal something of the politics of inevitability which emerged in this period.

This study has explored the interplay between broader trends in policy-making in the 1990s and the contingencies of policy-making in Berlin. Specifically, it has aimed to capture some of the ways in which neo-liberalism and new forms of governance were contingently produced in Berlin. It was argued that combining an analysis of the two would reveal something of the nature of "governance-beyond-the-state": the emergence of new formal and informal institutional arrangements of governance in the context of neo-liberalism (Swyngedouw 2005). The study has explored the ways in which such forms of governance have been informed by neo-liberalism and the ways in which they have, in turn, shaped the form neo-liberalism took in Berlin. The next Chapter will reflect on the merits of the framework employed to achieve this. The purpose of this Chapter is to offer some concluding remarks about the BWB case and to consider its wider significance.

The partial privatisation of BWB

Three inter-related themes characterised urban governance in 1990s Berlin and shaped the context in which BWB was partially privatised. First, there was the challenge of re-entering capitalist markets, of adapting to economic globalisation. In a sense, the context of urban governance bears some comparison to that of the countries of Central and Eastern Europe. However, the privatisation of BWB was not obviously the direct result of pressure from international organisations in Berlin, as it often was in Central and Eastern Europe (Castro 2009, 25). Privatisation and the embedding of neo-liberalism in Berlin occurred differently. This was, in part, because of the second theme which characterised policy-making: the notion that Berlin had the attributes to become a 'global city'.

Berlin was the capital of the newly reunified Germany, the largest economy in the EU. The city now had a history to catch-up with – to return to being one of the world's great cities. The context of policy-making and the expectations were different in Berlin, even if the neo-liberal policies were not. There was a widely held belief that, although Berlin was experiencing similar processes of socio-economic transition, it was strategically placed to benefit from the emerging markets in Central and Eastern Europe. To achieve this, Berlin had to conform to the neo-liberal norms of the global economy; through restructuring success would inevitably come.

There was something of a paradox encapsulated in the global city aspiration. On the one hand, Berlin was just re-entering capitalist markets, on the other, it was already becoming a global city. This was the great promise of neo-liberalism – the potency of the discourse and the story it told about Berlin's future. It provided the hope of quick prosperity. The final theme characterising policy-making in 1990s Berlin was that of reunification. This provided financial and technical challenges across the city such as the re-integration of infrastructures. It also provided much of what was unique to Berlin politics in the period.

The importance of these themes in water governance was clear. The interplay between them shaped the conditions in which the BWB privatisation occurred. The process of adapting to economic globalisation was evident in the commercialisation of some of BWB's functions in 1994. This was a trend apparent around the world as private and municipally-owned companies become competitors in global water markets (Swyngedouw 2003, 8). Most other major German cities took similar steps to change the legal status of their publicly owned companies to allow them to pursue commercial objectives outside of their borders (Wissen and Naumann 2006, 3). Unlike most other major German cities, however, privatisation followed commercialisation in Berlin. The partial privatisation of BWB was, in this sense, not typical within the German context of strong, albeit commercially-orientated municipalities. Why, then, did privatisation occur in Berlin and not in most other German cities?

Privatisation was more common in Eastern Germany. Rostock and Potsdam were the other main examples of privatisations in larger cities and private sector involvement was prominent throughout the smaller municipalities of Eastern Germany (see Naumann 2009). This suggests that privatisation emerged as a result of the economic problems caused by reunification and adjustments to capitalist markets. In the case of BWB, privatisation was 'inevitable' because of the fiscal crisis which emerged in the mid-1990s. This is the standard or ready-made explanation given for the privatisation by most interviewees and researchers (Fitch 2007a; Monstadt 2007) and the main argument put forward by the government.

Others (Pawlowski; Lederer) argued that the failure of BWB's commercial ventures between 1994-1999 made privatisation inevitable: that had this commercialisation been profitable privatisation would not have occurred (as it had not in most other German cities). Finally, some have highlighted the roles of 'ideological' actors like Fugmann-Heesing in implementing the privatisation and other neo-liberal policies in Berlin (*e.g.* Passadakis 2006; Werle 2004).

The global city policy discourse

This book has presented an alternative account of the privatisation, one which has stressed the contingency of political agency and the importance of the discursive effects of neo-liberalism. Fiscal pressures were apparent in the BWB debate. However, fiscal pressures do not inevitably lead to privatisation. For example, in Central and Eastern Europe in this period, despite the external pressure to privatise there was a trend towards maintaining and developing municipal water companies using finance from development banks in Poland, Hungary and Baltic States (Lobina 2001, 18). There were, then, alternatives to privatisation even in contexts of socio-economic transition. Furthermore, political actors are undoubtedly important but cannot always – at least in representative democracies – impose their will on politics. This study has argued that to understand the partial privatisation of BWB it is necessary to examine the discursive dimension in policy-making. Understanding policy-making as a contest over meanings, in which values inform the making of facts, has allowed the study to see this inevitability as a construct and the policy-making which surrounded it as the processes through which this 'truth' was produced. Neo-liberalism informed these processes of creating facts (e.g. that private sector management would be superior), while the claims to rationalism and objectivity underpinning the political system served to cloak the embedding of neo-liberal assumptions and values.

This study has analysed the BWB partial privatisation not with reference to the ready-made accounts of 'no alternatives' and 'necessities', but through a reflection on the diffuse ways in which neo-liberalism shaped policy-making in 1990s Berlin. The aim, instead, has been to explore how a lack of alternatives, supposed necessities of urban policy-making and inevitability were produced in Berlin. The 'necessity' of the partial privatisation was not, then, seen as emerging from a logical assessment of facts but was, rather, a construct arising from the embedding of a particular, neo-liberal way of seeing the 'facts'.

Policy-making is about making representations of the present (e.g. how BWB functioned as a public company) and the future (e.g. how it would in the future) authoritative. In accepting the discursive potency of neo-liberalism in this process it was also necessary to stress that neo-liberalism had to be produced by actors. Neo-liberalism may have shaped policy-making but the form it took in 1990s Berlin was, in turn, shaped by political agency in Berlin. It had to be embedded through policy-making. There have, then, been two strands to the analysis: a concern for the outcomes of political agency within Berlin, and the effects of the broader discourse of neo-liberalism in shaping this context of action. On the one hand, the study has been concerned with exploring the discursive work required to produce this inevitability. Discourse had material effects – it shaped policies in Berlin that did not achieve their objectives; that made the situation worse, not better. On the other hand, such policies were shaped and produced by actors – they were the result of decisions made through the political system. The study was, thus, fundamentally about how a decision-making process functions in the heat of a policy controversy such as water privatisation. It aimed to get into the thick of exchanges between actors, exploring how arguments for and against privatisation emerged, how these conflicts were articulated and how – through the institutions and processes of the political system – they ultimately produced the BWB partial privatisation.

From such a perspective the explanation for the privatisation lay somewhere between local and global dynamics and in both the constitutive effects of discourse and the political agency of actors on the ground producing neo-liberal policies. The embedding of neo-liberal assumptions and facts in the policy-making landscape set-up a discursive context in which privatisation could be presented by the government as inevitable in the context of the failure of the BWB commercialisation and the fiscal crisis of the 1990s. Thus the apparent necessity of the privatisation in 1997 was actually contingent upon a range of policy decisions and their effects over a long period of time. The privatisation was not simply the outcome of fiscal pressures *i.e.* a pragmatic response to fiscal pressures at the time.

Presented as a 'truth' and wrapped up in the success and prosperity it would bring, the global city policy discourse was rather a claim about how the future would be as Berlin adapted to economic globalisation. It was here that the storyline of inevitable socio-economic restructuring was vital to understanding the privatisation and policy-making more generally. It explained policy failures and provided solutions. It embedded a 'circular belief', in which the failure of one neo-liberal policy provided the government with the justification for another. Policies may have adapted and changed, but the rationality of neo-liberalism persevered. It was also used strategically to obscure the role of the government in the fiscal crisis as the mini-economic boom of the early 1990s receded. In these ways the global policy discourse acted as a key intellectual and normative means of interpreting reality – of creating knowledge and producing facts. It was fundamental to creating a politics of inevitability, in which there were no real alternatives to neo-liberal policies in the context of globalisation.

This pervading influence can be described as a neo-liberal mentality of governing. The notion of governmentality captures well this sense that certain problematisations and policy solutions can become 'normalised'. To become a global city, Berlin *had* to build offices for the companies who would come to the city. The boosterism of Berlin rested on the notion that Berlin was somehow on a set path which would lead to economic success. Berlin was always becoming (a global city) despite the realities which contradicted this. The storyline was both productive of, and a justification for, the contradiction between the image of the global city and socio-economic realities.

Through translations of the global city discourse into policy a neo-liberal governmentality was gradually embedded. Aspects of policy-making became black-boxed: there was a 'disappearance' of the uncertainty and alternatives to global city policies. Thus Berlin's aspiration to become a global city came to take on a commonsensical air. In this sense, the BWB privatisation policy-making process can in reality be extended to include the 'background' of 1990s policy-making. The importance of this preceding phase was seen as setting-up the problematisation of BWB, the point at which the official privatisation debate began. The official debate regarding BWB often failed to unpack that which had been black boxed earlier in the 1990s e.g. should BWB have aspired to become a 'global player'.

Importantly this neo-liberal discourse acted not merely as a means of interpreting policy-making problems and solutions in 1997, it also shaped, materially as well as discursively, the context of 1997. The prior translations of the policy discourse (the policies implemented in pursuit of the global city objective) had contributed to the fiscal crisis and the poor performance of BWB, the two main explanations given for the need to privatise BWB. The material and discursive context of the privatisation was then shaped by the unintended as well as the intended consequences of policy: by the failure of policies to make Berlin a prosperous global city and the failure of BWB to become a 'global player'.

There was not then a discernable long-term strategy to privatise BWB. In this sense it bears closer comparison to Maloney and Richardson's (1995) description of the privatisation of the regional water companies in England and Wales. Maloney and Richardson highlighted a mix of long-term structural trends and short-term political actions which led to the emergence of water privatisation. In the end, the privatisation of BWB arose not as the result of the government's ideological convictions alone but from the contingencies of policy-making in the period (119ff). Though there were certainly consultants pushing for privatisation, and members of the CDU had proposed it as early as 1992 (Lanz and Eitner 2005a, 10), there was no clear consensus in the governing coalition until 1997. An SPD member of the government stated that they had no plans to privatise BWB in 1996 (Interview: 12). The government appeared, at times, to be hesitant about the privatisation of BWB. Thus the privatisation emerged from a process of interpreting and re-interpreting the changing politico-economic context. It was the result of the complex interplay between discourse, economic conditions and political agency.

Ultimately, the case revealed that privatisation had little, if anything, to do with water management. It reveals the extent to which the traditional concerns of water policy-making and management were subordinated to the broader objectives of the government. Shifts towards neo-liberal practices of government led first to the re-making of BWB as a commercial enterprise and then as a set of assets which could and should be sold. In the process, the environmental and technical aspects management were, if not entirely neglected, certainly marginalised in the privatisation process. The need for private sector expertise was justified with reference to the claim that they would make BWB a more astute and profitable player in expanding international water markets. The emphasis was not on the inefficiency of the publicly managed BWB as it had been in the UK, for example. If anything, the argument was that the private sector would be better equipped to exploit BWB's existing expertise in water management.

New forms of governance, neo-liberalism and the partial privatisation

The privatisation was thus strategically produced in the context of a fiscal crisis and the failed commercialisation of BWB. Neo-liberalism had shaped the context in which the privatisation debate emerged and the apparent range of policy options available to actors. Still, the inevitability of the privatisation had to be produced through the exchanges between political actors: it had to be constructed. The government did not have one clear idea of the form the policy would take. A model of privatisation was not simply enforced through the policy process. Instead, the policy emerged through the policy debate and the implementation stage of the privatisation.

The detailed analysis of the policy process revealed the extent to which the partial privatisation (the Partial Privatisation Law and the contractual agreements with the private partners) was contingent upon the series of deals and compromises between the government and the SPD, the Unions and the Workers and the Parliament itself. Here the notion of translation – the simultaneous production of knowledge and creation of new relations – helped to expose the fragile foundations of the privatisation. It revealed a partial privatisation led most vociferously by the SPD leadership in the coalition government. By contrast, the CDU were revealed to be more circumspect and canny in their support, eventually brokering the deal to 'guarantee' job security and endanger the objective of a high sale price. A pivotal point in the process was when a slim majority in the SPD Parliamentary and Regional parties supported the privatisation, which led swiftly to a formal governmental proposal and the isolation of the opposition, most pertinently the Unions and BWB Workers.

The study argued that the embedding of neo-liberalism was inextricably linked to the extension of new forms of governance in policy-making in 1990s Berlin: formal and informal arrangements of public and private actors, in which consultants played prominent roles. These were apparent in the formal institutional arrangements of the Bankgesellschaft; the informal arrangements through which urban marketing firms promoted the New Berlin; the informal advisory roles played by consultants in the commercialisation of BWB; and the partial privatisation of BWB, as well as the ad hoc, extra-formal set of arrangements through which consultants and select public actors at the Finance Senate negotiated with the private sector companies.

In this sense the privatisation management team was emblematic of global city policy-making practices – indicative of a neo-liberal re-ordering of politics. As was argued at the end of Chapter 5, the global city policy discourse provided not just a means of thinking about policy-making (a neo-liberal way of thinking), but it also provided a means of making policy (a neo-liberal way of practising politics). As with the prior translations, the emergence of this public-private policy implementation team was presented as rational, normal and necessary to government. If the global city discourse revealed how broader discourses mixed with local conditions, an examination of the institutional processes of policy-making revealed how a neo-liberal rationality informed and was informed by exchanges between local actors and the structures of governance which emerged from these exchanges.

The emergence and influence of this bidding management team can be attributed to a number of factors. First, the institutional privileges of government: Finance was officially responsible for the privatisation. Fugmann-Heesing had the opportunity to shape the management of the process and she also had the strength of personality necessary to do this. Third, the consultants had knowledge and experience of privatisations not found in the rest of the political system: they were the designated experts. As important as these resources and institutional privileges were, their capacity to operate outside of the formal political process – to make decisions independently was crucial to their effectiveness. The institutional processes of policy-making were not determined by formal models. Instead, they were agreed informally, in an ad hoc manner. As a consequence, the management team were working in something of an "institutional void" (Hajer 2003b, 175). They operated in a new, extra-formal political space and were able to exploit the fact that they were not subject to the norms and rules of the formal political system. The strategic aim of the management team was to de-politicise the process: to reduce political debate and controversy and restrict flows of information to ensure a high sale price. As such, this form of governance was informed by a neo-liberal rationality. It was the key institutional form through which the political programme of privatisation was realised.

Within this extra-formal process, crucial aspects of not only the privatisation contracts, but also the Partial Privatisation Law were discussed and often decided upon. This attempt to make decisions apart from the rest of the political system was not entirely successful. Crucial parts of the final Partial Privatisation Law and contractual arrangements were influenced by actors in the wider process. Overall, power did not reside with one set of actors – not even the privatisation management team and the bidding companies. The exercise of political power was more contingent and unstable. Drawing on a variety of resources – the formal procedures of the political system, the official powers of political office and strategic use of the media – other actors were able, at various points, to exert an influence on the privatisation. Members of the opposition parties utilised formal legal mechanisms to challenge the privatisation, while traditional patterns of government – a generally corporatist approach – led to the inclusion of the Unions and Workers in at least some form of bargaining.

The ready-made accounts, alluding to the models of the classical modernist system (ready-made politics 1) and the managerialist government (ready-made politics 2) did not fully capture the ways in which the political system functioned. Key decision-making processes occurred in ad-hoc public-private structures alongside the official policy-making process. Consultants were crucial to this governance arrangement. Parliament was not simply a rubber-stamp for the policies of the executive but its scrutiny of the contracts and Partial Privatisation Law was limited intentionally by the management team. Overall, the BWB case reveals a policy process ultimately lacking in legitimacy, transparency and accountability.

Blurring the distinction between public and private

The partial privatisation policy process emphasises the contingencies of achieving privatisation and the need to avoid reifying neo-liberalism; to see it not as some omnipotent force working its way through politics, but rather as an outcome of very specific actions and unintended consequences of policy. This was not then a simple process of handing over power to the private sector. Governmentality studies characterise neo-liberalism as a "re-coding" of the state's position (Miller and Rose 2008, 80). Neo-liberal reforms of the state have led to less government but more governance (Larner 2000, 12). Fixed in the contract between the state of Berlin and the private sector companies, the position of the executive is, perhaps, in some ways stronger for being further removed from the rest of the traditional decision-making institutions of the state.

The BWB case was also reflective of the wider trend of governments actively facilitating privatisation. As Swyngedouw (2003, 11) states, governments have to provide incentives for the private companies, change laws, create new institutional frameworks, and agree conditions and compromises to ensure enough political support. There is then a contradiction between the stated aims and the means and forms through which privatisation is realised. Contradictions are apparent between the stated objective of reducing the role of government and the extent to which elements of the executive control was exerted (through the management team in particular) to realise the privatisation. A contradiction is also evident between the aim of creating a transparent, less centralised form of governance and the degree to which the final form of privatisation removed the executive and the private sector partners from parliamentary scrutiny. Furthermore, the extent to which public and private partners are locked together in decision-making can be seen to contradict the stated objective of bringing in private sector know-how to free up the company from the inefficiencies of public ownership and management.

Additionally, the parties to the privatisation process in 1999 agreed on informal rules and procedures which now shape aspects of management decision-making. This outcome can be directly linked to the way in which BWB was privatised. Certainly, the confidentiality of the bidding process is entirely unacceptable given the amount of key decisions made first there and then presented to the rest of the government and the Parliament. This was identified as the key problem with the process – the crucial deviation from the ready-made models of democratic decision-making. With the establishment of this parallel, extra-formal process, consultants became central actors in the realisation of the privatisation. They were integral to a new form of governance which served to undermine claims made by these official models to objectivity and legitimacy in policy-making.

Thus the Berlin case has revealed the high-level of dependencies between public and private actors in policy processes, with consultants operating at vital points of knowledge production and decision-making; they were crucial in shaping new forms of governing. As a result, knowledge, ways of doing things, of organizing the places where public and private meet, was increasingly defined by new forms of governance in which private sector actors such as consultants played a crucial role. In Berlin this meant consultancies such as Merrill Lynch. Though it is important to stress again that the policy process did provide certain checks and balances, the BWB case can, nonetheless, be seen as emblematic of the move towards an "increasingly remote and commercialized policy-making process" (Shore and Wright 1997, 3). A policy process in which decision-making was increasingly defined in the manner of a business deal, with private sector consultants at the centre of this re-ordering of the formal political system.

Though consultants were crucial to this new form of governance, the notion of 'consultocracy' is too simplistic a critique of the relationships between politicians, consultants and civil servants apparent in the BWB case. Those working within the government could be classified as 'hired guns' working for key governmental actors, they could also be seen as key knowledge providers and leaders of the process. Consultants were also lobbying government, promoting their own and other private sector interests. They were simply present at every stage and in every realm of the political process.

However, despite the undoubted importance of consultants it is necessary to stress the limits on their influence and the continuing importance of governmental figures. In being critical of their roles, there is a danger that their influence can be over-stated and their importance reified. In his detailed study of management consultants in Britain, Canada and France, Saint-Martin (2000, 205) identified four key issues relevant to an assessment of their effects on democracy. Though his study focused less on policy-making processes and more on the management of bureaucracies, they provide a useful means of organising concluding remarks on consultants and the BWB case.

Saint-Martin's (2000) first concern was this issue of 'consultocracy': whether consultants were increasingly acquiring the powers assigned to elected officials and appointed bureaucrats through the re-definition of government in managerial terms (206-6). The second issue was whether the reliance on consultants in policy-making was having negative effects on the transparency and openness of debates (206). Third, was a more general point about the neutrality or independence as policy experts (206): given their business interests their knowledge is not merely problem-solving; it is also a product to make money (206). The final issue was also related to the fact that they were first and foremost commercial actors – 'conflicts of interest' (208).

On the first point, the Berlin case certainly reveals the extent to which a wider range of interests were instrumental in the designing of policy. The management team at Finance was certainly highly influential, but it was unable to dominate the legislative and judicial processes. As stated, research was unable to determine the exact influence of consultants (and private interests more generally) on the decision to privatise BWB. It is, however, clear that consultants were involved and it is clear that politicians wanted them to be. Opposition Party politicians stated that the politicians went looking for the consultants during the 1990s (Interviews: 6, 7). Certainly consultants were aiming for, and achieving, influence in the process. They were indeed integral to global city policy-making; but to whose benefit? They definitely made their companies wealthier but as the SPD parliamentarian stated, the consultants working for Fugmann-Heesing were also "enriching the power of (the Senate of) Finance" (Interview: 16). There was an alignment of interests in this new form of governance.

On the second issue of transparency, the privatisation process was, despite the resistance from other actors, characterised by a lack of openness. As such the process was, at least from the point of 'implementation' onwards, problematic. On the third point regarding their interests and independence, their knowledge was indeed a product and the privatisation process something of a business opportunity for some e.g. the consultant, Pröpper sold his knowledge of privatisation to both RWE and the Berlin government. Finally, the high number of conflicts of interest which emerged unquestionably raises concerns about the accountability and legitimacy of a political process populated by consultants. The involvement of so many private sector actors (the companies as well as the consultants) created many moments of unease and suspicion in the privatisation. Even if no clear case of corruption has been unearthed, the involvement of consultants increased the potential opportunities for corruption to occur.

Experts, contemporary governance and neo-liberalism

Overall, then it is possible to discern a number of problematic issues in the BWB case concerning the role of consultants. Particularly, it raises questions about the extent to which policy-making processes are detached from citizens and even elected politicians. In this sense the BWB case is not that remarkable, but emblematic of a more general trend in representative democracies. The concern over the role of experts in politics has led Fischer to state:

> "The institutions and practices of modern societies are anchored to professional knowledge, leaving decision-makers highly dependent on expert judgements. Politicians and policy decision-makers, along with citizens, are left to depend on and trust the validity of the knowledge and competences of the experts who make them" (Fischer 2009, 3).

This dominance of expertise in modern democracies rests upon a more fundamental characteristic of political systems. Expertise – of whatever form – is reinforced, enabled, by a claim to objectivity and rationalism: the knowledge they produce is presented as neutral, "independent of time and context" and given the complexity of contemporary political problems there is little option but to defer to their judgements (Fischer 2009, 3). Of course, there is no reason to assume that such advice is neutral or that experts are able to make better decisions than anyone else. One aim of this study has been to highlight the way in which neo-liberal assumptions and values shaped the construction of facts in Berlin politics. It has problematised the neutrality and rationalism of political decision-making.

This is why reifications of models of political decision-making, as alluded to in Chapter 6, are problematic: they seek to distract critical eyes from the actual processes of governing. In this sense the models of democratic decision-making and the orthodoxy on which they rest (the notions of neutral, technical-rational, bureaucratic decision-making), come to have the potency of "legitimizing myth" (Miller and Fox 2007, 3). They allow actors a ready-made image of policy-making and implementation they can present to the rest of the world. This study briefly touched upon the traditional hierarchical form of government (in ready-made politics 1). The main focus, however, was new forms of governance and the new managerialism, which came to be a characteristic feature of policy-making in Berlin and elsewhere. Such notions justify the role of consultants in governing (and were alluded to in ready-made politics 2). The study pointed to some of the problems inherent to such shifts from government to governance. As such it problematised the tendency to assume that such new forms of governance are an improvement upon government (Mayntz 2003, 32).

Such findings chime with the work of other researchers of contemporary policy-making and public administration. In their critique of US politics, Miller and Fox (2007) saw increasing evidence that democracy and policy-making could no longer be said to be fully captured by the framework of classical-modernist institutions. They (Miller and Fox 2007, 6) argue that this formal model with its notions of the 'public interest' or 'common good', can, in fact, as the BWB case reveals, work to conceal the influence of particularistic actors' on government who conduct policy-making "outside of the limelight". These are spaces in which "knowledgeable" actors define, clarify and resolve policy problems through the dissemination and discussion of information they deem relevant (Miller and Fox 2007, 6). Tucked away from the formal processes of the model, policy-making and implementation are often dominated not only by official actors such as civil servants and politicians but also by "policy implementers" (Miller and Fox 2007, 6). In the BWB case, consultants fulfilled these roles. Their specialist knowledge of privatisations provided them with a key resource in defining the privatisation process. As the BWB case reveals, the presence of these extra formal arrangements often undermine the very aims of the model: to provide transparency, accountability and legitimacy.

The arrangement of public and private actors in such new forms of governance thus raises concerns about the health of contemporary democracies. This study has argued that the emergence of such problematic forms of governance should be understood within the broader context of neo-liberalism: that governance should be seen as neo-liberal governmentality. From a governmentality perspective, neo-liberalism is not just a set of policies like privatisation. Rather it is, to return again to Brown (2005, 38), also "a political rationality that both organizes these policies and reaches beyond the market" (Brown 2005, 38). This study has argued that a neo-liberal rationality shaped the production of public-private partnerships in governance and provided the *raison d'être* for a politics of necessities and no alternatives.

It should be stressed, however, that the problems arising from the use of consultants and the more general re-definition of governing of which they are a part, should not be seen as arising only from neo-liberalism. Neo-liberalism continues to depend upon and even extends further the notion of instrumental rationality which has long underpinned the workings of representative democracies (Miller and Fox 2007, 38). The BWB case has revealed that neo-liberal reforms did not overhaul the political system. There was a range of rationalities and institutions at work and in contest with each other. This is in part why varying accounts of the partial privatisation process emerged in interviews. Nonetheless, the attempted de-politicisation of the policy process, the prominence of consultants and the privatisation management team, are indicative of the broader condition of "post-politics" (Swyngedouw 2007a; 2009a). Their centrality to policy-making is representative of the belief that what is required in contemporary politics, in systems of 'good governance', is the expert management of public-private arrangements.

8. Assessing the theoretical approach

Introduction

Although only the BWB case was analysed, the approach employed here was not specifically oriented to water governance, privatisation processes or to the context of Berlin in the 1990s. Rather the aim was to develop a general, constructivist approach to policy-making, one which addressed the interdependencies between discourse, political agency and political institutions in the making of policy. In doing so, the hope was that the approach would add to the growing post-positivist policy studies literature. This final chapter offers an assessment of this approach, considering in particular the benefits gained from this synthesis, the questions that remained unanswered and its potential for development in the future.

Synthesising three literatures

The point of this synthesis of literatures was not to produce an overarching conceptual framework, one outlining a structured means through which research should be conducted and phenomenon analysed and understood. Instead, the synthesis was exploratory, revealing complementary and contrasting assumptions about the world. From this a series of questions and concerns emerged to guide the research. There was, of course, a logic in bringing the three literatures together: all three focus attention on the production of facts. By drawing on aspects of each of the literatures, balancing oversights with insights, the purpose of the approach was to address the challenges of researching global-local dynamics and structure-agency issues in the making of policy.

Starting from the assumption common to the post-positivist literature – that policy is a construct – knowledge about the world was seen as being partially and contingently produced. From such a perspective there remained a separate reality beyond construction, but 'facts' had to be produced and this was a process in which assumptions and values played a role. Problematising claims to objectivity and rationality, the aim was to explore how meanings were produced in politics and importantly, how, they became rationalised.

It was argued that the policy process should be seen not merely as a discursive process but one ultimately contingent upon the strategic exchanges between actors in policy processes. The study, in the attempt to conceptualise these processes, centred on a tension between discourse and agency. The aim was to balance an appreciation of the constitutive effects of discourse in policy-making, with an understanding that actors are, in the end, the makers of policy. Thus though the approach may have been post-structuralist in orientation it had an almost realist concern to reveal the institutional processes through which discourses were produced.

To productively accommodate this tension, the approach brought together elements of three literatures. To tackle the discursive dimension of policy-making, the study drew on Neo-Foucauldian governmentality studies, particularly the work of Rose and Miller (1992) and, from the field of interpretivist policy studies, Hajer's work on policy discourses. To think about the way in which local-level interactions are productive of policies, discourses and the embedding of governmentalities, the notion of translation from ANT was utilised. Additionally, this concern with actors, institutions and the intricacies of policy processes led to further engagement with the interpretivist policy studies literature, particularly that critiquing the ways in which democratic institutions functioned.

The following sections assess the ways in which the approach was operationalised to tackle the key research aims outlined in Chapter 1: the need to problematise the production of facts in policy-making, to develop an understanding of the interactions between global and local dynamics and discourse and political agency.

Problematising the production of facts: ready-made versus politics in the making

ANT provided a useful way of characterising this work. Drawing on Latour (1987) a distinction was made between 'ready-made politics' and 'politics in the making' accounts of the policy process. The former captured the way in which actors – wittingly and unwittingly – rationalised decision-making processes with hindsight. Most pertinently, how they 'black-boxed' the policy process in the 'objectivism' and 'rationality' of the formal institutional processes, with their claims to democratic decision-making. Ultimately, the politics in the making account made no claims to being an authoritative account. It too was based on actors' interpretations of the process as well as this author's interpretation of their accounts. It was then an incomplete account, one with its own epistemological assumptions. It was also heavily dependent upon interviewees, who undoubtedly had their own agendas and unique memories of the process. Nonetheless, through bringing a range of interpretations together and drawing on media and other secondary sources, the politics in the making account provided a nuanced account of a complicated, problematic policy process.

This distinction between ready-made politics and politics in the making provided a means through which the analysis could move beyond a simple acceptance of formal images (of the classical-modernist system or the managerialist account), as alluded to by participants, whilst at the same time considering their significance to the policy process. In particular, it was possible to consider how such images of policy-making could be strategically deployed by actors to conceal more problematic processes. When combined with insights from the post-positivist policy studies literature it was possible to critique such models. Using this approach, the depoliticised accounts of the BWB privatisation (the ready-made versions) were undermined.

Global-local dynamics

The basis of the approach was that policy-making should be placed within a broader understanding of governance as governmentality. Most importantly, it brought attention to the inter-dependencies between thought and practices of government: the connections between claims to what is 'good' and what can be practically realised in politics (Rose and Miller 1992, 181-2). Adopting such an approach does risk underplaying material structures – particularly of economic processes. Certainly, this study has not provided a detailed account of economic and financial conditions in Berlin during the 1990s. It is hoped, however, that the – intended and unintended – material effects of discourses were incorporated in the analysis of the global city discourse (Chapter 5). From a governmentality perspective policy-making is the embedding of particular normative and theoretical assumptions in the discursive and material practices of government.

Such an approach emphasised that governing was contingent upon knowledge production. Particular forms of expertise were essential to providing the means of representing and governing reality. This study has argued that private sector consultants should be seen in these terms, as embodying an expertise essential to neo-liberal governmentalities; their knowledge of the practices of the business world integral to the privatisation process.

Rose and Miller's work provided a general means of understanding the global-local dynamics at play in policy-making. They argued that the interplay between broader political rationalities and local governmental practices should be understood as one of 'translation': as a process of transference and transformation. They do not, however, elaborate upon this process: their approach does not provide specific conceptual tools to expose this interplay, which, in a detailed analysis of policy, is necessary. To better explore the discursive context and the ways in which broader discourses are localised, Hajer's conceptualisation of policy discourses was introduced.

Policy discourses: political rationalities made local, contingent

The "terms" of policy discourses, "narrative storylines", "policy vocabularies" and "generative metaphors" Hajer 2003a, 103) shape the production of meanings in policy-making (Hajer 2003a, 107). Tracing the terms of policy discourses emphasised the more local intellectual and normative processes through which meaning was produced in policy-making. In the BWB case study, Hajer's notion of policy discourses revealed how the assumptions and strategies of the broader discourse of neo-liberalism became embedded in the local conditions of 1990s Berlin; it revealed how a "local" neo-liberalism (Peck and Tickell 2002, 46) emerged in Berlin.

This mid-range conceptual tool highlighted the importance of language in policy debates. Most particularly, the focus on policy discourse – the terms of the global city policy discourse, rather than simply on neo-liberalism – emphasised that water policy-making in Berlin was rarely explicitly 'neo-liberal'. Rather, neo-liberal assumptions and strategies were buried within the storyline of inevitable restructuring. In this way, searching for and analysing policy discourses provided a useful means of capturing the mixing of global and local dynamics. It provided an extra analytical layer. With its terms of discourse as analytical categories, the concept offered a relatively straightforward set of tools with which to unpack discourses.

Discourse and agency

Storylines, and the discourses which they animate and hold together, have to be produced. It is here that the issue of agency versus structure becomes most evident: how do policy discourses become a potent means of interpreting the policy-making landscape? A criticism of the governmentality literature is the inclination to concentrate on the discursive dimension of politics at the expense of the actual practices which both enable and are enabled by discourse. There is, then, a risk that such an approach can reify the potency of discourses (McKee 2009, 473). Indeed, a more general criticism about neo-Foucauldian discourse approaches is that they reveal very little about political agency and the processes through which interests and policy-making agendas emerge (Rutland and Aylett 2008, 631). This favouring of the discursive dimension of governing is perhaps understandable given the intended focus of the literature on unearthing mentalities of governing in societal practices. Nonetheless, questions remain unanswered regarding the interplay between discourse-agency. How far do actors shape, produce and re-produce discourses? How do broader political rationalities take root in local contexts of actions? What comes first, the discourse or agency?

This study does not, of course, claim to have found answers to these questions. Instead, it adopted a more practical approach. If it is impossible to truly analyse the direct interconnections between discourse and agency in any given context then some form of compromise must be found. The aim was to conduct a differentiated examination of both the discursive and institutional dimensions of the policy process, one at least suggestive of the interplay between agency and discourse. This was, perhaps, rather schematically done, but this should be seen as symptomatic of the difficulty of coming to terms with one of the key challenges in the social sciences.

The approach adopted here assumed that discourse is crucial to the making of policy, but that discourses themselves must be made. The overall shape and significance of policy discourses is beyond the grasp of those within its boundaries. In their entirety, they cannot be seen as intentionally produced, but they can serve strategic purposes. In providing a particular way of interpreting reality – normalising assumptions and strategies of political rationalities – they play to particular interests whilst marginalising others (Freeman 2007, 2). They are not stable because they are constantly being re-defined or rejected by actors. Discourses are thus neither quite the instruments utilised by actors nor the unseen, ungraspable force determining what they do. There is interplay between the two. Hence why it is necessary, alongside an examination of policy discourses, to examine political agency within the institutional context of the policy process.

As one of the key writers in the field states, a governmentality perspective can reveal much about the rationalities of governing, but it cannot provide a complete account of politics (Dean 2008, 198). Specifically, the governmentality perspective does not accommodate a thorough analysis of the strategic interplay between actors: the "struggle or competition between competing forces, groups or individuals attempting to influence, appropriate or otherwise control the exercise of authority" (Dean 2008, 198).

Translation

To address this, the approached developed here placed a greater emphasis on the notion of translation, so as to allow for a more thorough examination of political agency and the formation of political priorities in policy-making (Rutland and Aylett 2008, 631). The obvious question here was why choose to continue with the notion of translation, a concept developed in the field of STS, when there is a vast choice of concepts within political science. Why not, for example, simply use Hajer's (1997) broader approach of "discourse coalitions" to examine the interplay between discourse and institutional processes? The main justification for employing elements of the notion of translation was that it shifted the line of analysis firmly away from discourse to the exchanges between actors: the emphasis in the analysis shifted from an, at times, almost "top-down" perspective to one which was distinctly "bottom-up" (Kendall 2004, 64). Though largely employed to analyse innovation, translation can be seen in more general terms as a means of understanding how actors negotiate new meanings and, in the process, the relationships upon which those meanings are contingent. It is therefore helpful for an analysis of the policy process.

The sociology of translation, as outlined by Callon (1986), was not utilised as a detailed analytical approach. Rather it was used as a more general way of thinking about the dynamics of the policy process. Understanding policy-making as entailing the re-representation and re-assembling (Freeman 2007, 12) of governmental practices – according to particular rationalities – was suggestive of the contingency of knowledge production upon the configuration of actors. Translation, as understood here, was as much a methodological guide as a conceptual insight. The aim was to produce a 'thick description' of the processes through which the privatisation was actually agreed by actors. The stress on the actors' accounts provided a better balance between discourse and agency. It also brought the analysis down to the level of Berlin politics. As such translation directs the analysis to the localised interactions through which policies, discourses and ultimately governmentalities are produced.

From a translation perspective agency is 'won', it is achieved through the building of relations and the agreement of meanings between actors. Agency is contingent; it is relationally produced. In the study of policy-making processes this means that political agency and political power should not be assumed. For example, policy should not be seen simply in terms of the executive acting on their pre-defined interests, imposing their will on the rest of the political institutions. Institutions have formal powers but placing these cause-effect mechanisms within a translation perspective means that roles, responsibilities and power must still be achieved in the policy process. The institutions of political systems are thus translated through policy-making.

Recently, other researchers (Rutland and Aylett 2008) have also argued that a synthesis of governmentality and ANT is useful for the analysis of policy-making. Their interest in translation does, however, hinge more on the inclusion of non-human entities in the analysis of policy and less on translation as a means of probing the interplay between discourse and agency. The move made in this study was comparable to that proposed by social policy researchers (Stenson 2005, 2008 – see McKee 2009) who have attempted to develop a "realist governmentality" approach in order to conduct detailed empirical research. In a similar vein, their argument is that a concern for micro-level actions must complement an analysis of the discursive field of governing (McKee 2009). One contribution of this book is that it combines a very detailed analysis of the interactions between actors in a policy process with a consideration of the broader rationalities and local policy discourses which shaped politics in the 1990s.

Areas for future research

The overall purpose of this approach is to provide a general means of critiquing policy-making. It is non-deterministic and general in its applicability. It does not provide a model of policy-making and it was not specifically designed for the study of privatisation. Instead, the approach offers a set of analytical concerns centred on the contingent production of facts and governance arrangements in policy-making. It draws together a range of complementary conceptual tools with which to address these concerns, exploring the interplays between political rationalities, expertise and political institutions. In the process, the approach touches upon the fundamental debates about the interdependencies between macro-micro processes and structure-agency in politics and provides a differentiated perspective on them. There is much scope for further development of such a synthetic approach to policy-making and politics in general. The following sections consider some of these, highlighting areas where this study might have gone further: examining consultants and governance, analysing policy-making as translation and problematising the 'post-political'. In the process it refers to literatures, their different concerns and questions, which might provide a means of refining the approach.

Consultants and contemporary governance

This study has revealed the integral role played by a range of private sector consultants in water policy-making and policy-making more generally. Given the evidence that the growth of consultants in policy-making and administration is likely to continue in the coming years, there is certainly a need for research on the scope of their work and their influence on policy-making (particularly as Saint-Martin (2000) is the only substantial study of consultants in politics). In Berlin, for example, the government has continued to employ consultants at great cost. For the privatisation of Berliner Sparkasse in 2007, the Finance Senate again assembled a team of consultants, this time consisting of the legal firm Freshfields, Bruckhaus & Deringer, the financial consultants UBS and the accountancy firm SUSAT (Der Tagesspiegel 2009).

Though this study made no attempt to develop a theoretical understanding of the work of private sector consultants, there is potential to do this with the approach employed. Here the work on experts and environmental policy-making from an STS background could provide a guide. In particular the notion of "co-production" as outlined by Jasanoff (2006) has potential synergies with the approach. This term was originally developed with regard to the interdependencies between scientific knowledge and social order (e.g. institutions, discourses and identities). The "co-production" perspective aims to reveal how technical knowledge "both embeds and is embedded in social practices, identities, norms, conventions, discourses, instruments and institutions – in short, in all the building blocks of what we term the social" (Jasanoff 2006, 3). It is compatible with the emphasis of the governmentality literature on the importance of expertise and the production of knowledge to governance. In its allusion to the simultaneity of knowledge production and the ordering of relations between things, it has obvious similarities to ANT. Indeed, ANT is a key influence on those writing in the "idiom" of co-production (Jasanoff 2006, 22). There is a shared emphasis in all the approaches on the partiality of knowledge production and the way in which this produces particular ways of representing and ordering reality.

In the field of co-production studies there has been interesting work focusing on the role of scientific experts in representing the environment and how, in turn, this has shaped the politics and governance of the environment. A recent example would be the roles played by computer modellers in making climate change 'real' and therefore governable (Edwards 2001 and the edited volume more generally, Miller and Edwards 2001). As Jasanoff (2006) writes "the ways in which we know and represent the world (both nature and society) are inseparable from the ways in which we choose to live in it" (Jasanoff 2006, 2). A more detailed examination of the knowledge that consultants produce in politics – along similar lines – might provide one way of assessing their importance to policy-making processes.

One obvious avenue for research would be to consider the re-configuration of environmental governance through the introduction of market-based forms of regulation. New Environmental Policy Instruments (NEPI) such as tradable permit schemes, voluntary agreements and environmental management systems have been interpreted as emblematic of the move from government to governance (Jordan *et al.* 2005). Seen from a governmentality perspective, they could be interpreted as neo-liberal governmental technologies dependent on, and perhaps enabling, new forms of expertise in environmental governance such as that provided by private sector consultants. Future research might look at the role consultants are playing in depicting new societal-environmental relations through the development of these market-based technologies of government. For example, what role are consultants playing in the management of carbon-trading emissions schemes? How dependent are biodiversity banking schemes on financial and economic consultants? Are such experts becoming as important to environmental governance as biologists or computer modellers? How far are they co-producing new neo-liberal environmental governmentalities? If so, what are the implications for the legitimacy of environmental policy-making?

Translation, governmentality and policy studies

This study made a preliminary attempt to think about policy-making processes in the general terms of translation. Future research might attempt a more thorough utilisation of the concept. In ANT, knowledge is constructed through translations and it is here that the concept has wider applicability in the social sciences: contests over 'facts', the creation of knowledge, are at the centre of controversies between actors (Garrety 1997, 755). Apart from the arguably problematic inclusion of non-human entities in the analysis, there is nothing intrinsic to the approach provided by Callon (1986) which prevents it from being applied in the field of policy-making. This early piece provides the clearest articulation of translation as a approach for conducting research. His four moments of translation – *problematisation*, *interessement*, *enrolment* and *mobilisation* – potentially provide a methodological approach and conceptual vocabulary with which to describe and analyse the micro-interactions of policy implementation. It could be particularly useful in revealing the contingency of policy implementation on an array of actors and the extent to which it is translated as it works its way through levels and sectors.

Given ANT's roots in anthropology, it might also be utilised to conduct an "anthropology of policy" (Wedel *et al.* 2005). Certainly, the attention to the terms of discourse in the approach outlined in this study could usefully address Shore and Wright's (1997) concern to explore the "mobilizing metaphors and linguistic devises that cloak policy with the symbols and trappings of political legitimacy" (3). Such a detailed, anthropological approach was not possible in the BWB case, given the time which has elapsed since the privatisation and the unwillingness of some actors to be interviewed (due to the controversy the privatisation has provoked). However, there is certainly potential for it to be applied in ongoing and recent cases where there is the possibility for participant observation or more detailed re-construction through interviews.

An obvious area in which translation could be utilised is that of policy transfer and policy change. Evocative of notions of transference and transformation (Czarniawska and Sevon 2005), translation could be utilised to provide accounts of how policies such as privatisation spread, moving from one sector to another and how, in the process, policies themselves change, come to mean different things as they are implemented. Possible synergies are apparent between the approach outlined in the study and the work of Diane Stone (1999, 55 quoted in Grin and Loeber 2007, 204), who has conceived of policy transfer as entailing the movement (and transformation) of "ideas, ideals, expertise, programmes and personnel". In other words, it is not merely policies which are being transferred but also knowledge, sets of relations and often actors.

Researching post-political policy-making

The key features of the BWB privatisation (the depoliticisation of decision-making and the problematic merging of public and private interests) are symptomatic of what a number of critics have identified as a more fundamental shift towards a post-political or post-democratic form of politics in the last few decades. The approach developed in this study, with its concern for the political construction of facts, has the potential to be utilised and refined for future research on this form of politics. To conclude this study it is appropriate to briefly consider what shape such research might take.

At present, most of the work in this field has been at the theoretical or general level: by political philosophers such as Chantal Mouffe, Jacques Rancierre and Slavoj Žižek, political scientists offering general accounts of contemporary politics (Crouch 2004; Hay 2007) or theory-driven geographers of the city and the environment (Swyngedouw 2007a; 2009a; 2010). There is, then, a need to provide more detailed, empirical accounts of post-politics; to expose the actualities of producing post-politics in specific contexts and explore the usefulness of the notion for analysing contemporary forms of governance.

In the current context of the global financial crisis, such work may become even more important given, the apparent lack of concerted opposition and alternatives in the political mainstream. There is a need to understand the continued potency of neo-liberalism despite the crisis of the global financial system it shaped. There is a need to understand why the Left's hope that this would lead to a "Socialist Moment" (Rawnsley 2010) was so short-lived. Future research on policy-making in post-political systems needs to problematise the continuing presentation of politics in apolitical terms and reveal in detail how apparent consensus is constructed and maintained. The 'politics in the making' approach employed here, with its focus on the production of facts and truths in policy-making, could be developed to explore the inter-relationships between expert-dominated policy processes, 'shadow' governance arrangements and the presentation of policies in neutral or technocratic terms.

The approach employed here provided a set of tools with which to problematise a politics of inevitability and destabilise a 'local' neo-liberalism. The analysis of the BWB privatisation presented revealed the range of deals and compromises required to realise neo-liberal policies, and the extent to which these neo-liberal policies were reformed in the process. It threw a light on the partial production of facts. In short, it did not reify the 'necessities' of policy-making in a context of neo-liberal globalisation. It picked apart the no alternative to privatisation argument to reveal a "deliberate policy of depoliticisation"; one common to neo-liberal reforms around the world (Munck 2005, 68).

Without wishing to underestimate the extent to which it is materially and discursively embedded in political systems, adopting such an approach to neo-liberalism makes it contestable in the local forms it takes. As stated in Chapter 3, political institutions and the logics which inform them are constructs that can be resisted and reformed. From such a perspective there are always alternatives. To say that there are not, is to – wittingly or unwittingly – accept a range of normative assumptions about the world, the means by which society should be governed and the way in which politics should be conducted.

References

Amt für Statistik Berlin-Brandenburg (2007) *Bevölkerungsstand in Berlin seit 1991*, Pressemitteilung vom 24.10.2007-Nr.249, Available: http://www.statistik-berlin-brandenburg.de/pms/2007/07-10-24a.pdf (10 December 2009).

Bakker, K. (2003a) 'From public to private to…mutual? Restructuring water supply governance in England and Wales', *Geoforum*, vol. 34, no. 3, pp. 359-374.

Bakker, K. (2003b) *An uncooperative commodity: privatising water in England and Wales*, Oxford: Oxford University Press.

Barraque, B. (2009) 'The development of water services in Europe: from diversity to convergence', in Castro, J. E. and Heller, L. (eds.) *Water and Sanitation Services: Public Policy and Management*, London: Earthscan.

Barry, A. (2001) *Political Machines: Governing a Technological Society*, London: Athlone.

Barry, A., Osborne, T. and Rose, N. (1996) (eds.) *Foucault and Political Reason: Liberalism, Neo-liberalism and Rationalities of Government*, London: Routledge Press.

Barzelay, M. (2001) *The New Public Management: Improving Research and Policy Dialogue*, Berkeley: University of California Press.

Baumgartner, F. R. and Jones, B. D. (1993) *Agendas and instability in American politics*, Chicago: University of Chicago Press.

Blair, T. and Schröder, G. (1999) *Europe: The Third Way / Die Neue Mitte*, London: The Labour Party.

Beaverstock, J. V., Smith, R. G. and Taylor, P. J. (1999) 'A Roster of World Cities', *Cities*, vol. 16, no. 6, December, pp. 445-458.

Bel, G. (2006) 'Retrospectives: The coining of Privatisation and Germany's National Socialist Party', *The Journal of Economic Perspectives*, vol. 20, no. 3, pp. 187-194.

Berliner Morgenpost (1997) 'Senat will eine Milliarde von den Wasserbetrieben', 15 July.

Berliner Morgenpost (1998) 'Pieroths Pläne sind sehr dubios', 25 May.

Berliner Morgenpost (1999) 'CDU will geplante Teilprivatisierung der Berliner Wasser-Betriebe stoppen', 13 March.
Berliner Morgenpost (1999b) 'Fugmann-Heesing hält fest an Merrill-Lynch', 24 March.

Berliner Zeitung (1997a) 'Wasser-Betriebe vor Verkauf?', 17 May.

Berliner Zeitung (1997b) 'Finanzsenatorin will alle Rathäuser verkaufen', 12 June.

Berliner Zeitung (1997c) 'BWB soll an die Börse', 14 July.

Berliner Zeitung (1997d) 'Bei den Wasserbetrieben geht die Angst um', 13 August.

Berliner Zeitung (1998) 'Verkauf der Wasserbetriebe erst im März 1999', 29 August.

Berliner Zeitung (1999a) 'Koaliton gibt grünes licht für BWB-Verkauf', 30 March.

Berliner Zeitung (1999b) 'Beim Zuschlag für Enron droht Anfechtung', 31 March.

Berliner Zeitung (1999c) 'Teilprivatisierung der Wasserbetriebe vollzogen', 30 October.

Beveridge, R. and Hueesker, F. (2007) 'Corruption, privatisation and democracy: The sale of Berlin's Water Company in 1999', *Corruption and Democracy in Europe: Public Opinion and Social Representations: ECPR General Conference*, Pisa, Italy, 6-8 September 2007.

Beveridge, R. and Pflug, T. (2007) 'The twists and turns of water management in Berlin and London', *5th International Water History Association (IWHA) Conference: Pasts and Futures of Water*, 13-17 June 2007.

Bevir, M. (2006) *Interpretive Approaches to British Government and Politics*, Multi-Campus (Online), Available: http://escholarship.org/uc/item/6hw0g1vc (26 July 2009).

Bevir, M. and Rhodes, R. (2006) 'Interpretive Approaches to British Governance and Politics', *British Politics*, vol. 1, no. 1, April, pp. 84-112.

Beyer, J. and Hopner, M. (2003) 'The disintegration of organised capitalism: German corporate governance in the 1990s', *West European Politics*, vol. 26, no. 4, pp. 179-198.

B.Z. (1997) 'Berlin-Auktion geht weiter', 21 May.

Bloor, D. (1999) 'Anti-Latour', *Studies In History and Philosophy of Science Part A*, vol. 30, no. 1, March, pp. 81-112.

Böhm, T. and Hasse, T. (1999b) 'Berlin wird modernste Stadt der Welt', *B.Z.* 1 March.

Bollmann, R. (1999) 'Koalitionskrach um Wasserbetriebe eskaliert', *Die Tageszeitung*, 20 May.

Brown, W. (2005) *Edgework: critical essays on knowledge and politics*, Princeton: Princeton University Press.

Brügel, I. (1993) 'Local economic development in the transformation of Berlin', *Regional Studies*, vol. 27, no. 2, pp. 155-159.

Callon, M. (1986) 'Some elements of a sociology of translation: domestication of the scallops and the fishermen of St Brieuc Bay', in Law, J. (ed.) *Power, Action and Belief: A new sociology of knowledge? Sociological Review Monograph 32*, London: Routledge and Kegan Paul.

Callon, M. (1991) 'Techno-economic networks and irreversibility', in Law, J. (ed.) *A Sociology of Monsters: Essays on Power, Technology and Domination, Sociological Review Monograph 38*, London: Routledge Press.

Callon, M. (1999) 'Actor-Network Theory: the Market Test', in Law, J. and Hassard, J. (eds.) *Actor Network and After*, Oxford and Keele: Blackwell and the Sociological Review.

Callon, M. and Latour, B. (1981) 'Unscrewing the big Leviathan: how actors macrostructure reality and how sociologists help them do so', in Knorr-Cetina, K. and Cicourel, A. V. (eds.) *Advances in Social Theory and Methodology*, London: Routledge and Keegan Paul.

Callon, M., Meadel, C. and Rabeharisoa, V. (2002) 'The economy of qualities', *Economy and Society*, vol. 31, no. 2, pp. 194-217.

Campbell, S. (1999) 'Capital reconstruction and capital accumulation in Berlin: a reply to Peter Marcuse', *International Journal of Urban and Regional Research*, vol. 23, no. 1, pp. 173-179.

Carver, T. (2002) 'Discourse Analysis and the "Linguistic Turn"', *European Political Science*, Autumn, vol. 2, no.1 (Online), Available: http://www.essex.ac.uk/ECPR/publications/eps/onlineissues/ autumn2002/research/carver.htm. (30 April 2010)

Castro, J. E. (2002) 'Arguments underlying current programmes promoting private participation in water and sanitation services', in Castro, J. E. (Coord.), *PRINWASS Project*, European Commission - 5th Framework Programme, INCO-DEV, Contract ICA4-CT-2001-10041, University of Oxford.

Castro, J. E. (2006) 'Urban conflicts over water in Mexico: a theoretical and empirical exploration', in Barraqué, B. (ed.) *Urban Water Conflicts: an Analysis on the Origins and Nature of Water-related Unrest and Conflicts in the Urban Setting*, Paris: UNESCO Working series SC-2006/WS/19 (2006).

Castro, J. E. (2009) '*Introduction*', in Castro, J. E. and Heller, L. (eds.), *Water and Sanitation Services: Public Policy and Management*, London and Sterling, VA: Earthscan.

Castro, J. E., Kaika, M. and Swyngedouw, E. (2003) 'London: structural continuities and institutional change in water management', *European Planning Studies*, Special issue on Water for the city: trends, policy issues and the challenge of sustainability, vol. 11, no. 3, pp. 283-298.

Centeno, M. A. (2001) 'Isomorphic neoliberalism and the creation of inevitability', *SASE Conference Amsterdam*, (Online) Available: www.princeton.edu/~cenmiga/works/SASE%20Centeno.doc (20 March 2010).

Chacon, B. (2005) *Die Beeinflussung politischer Entscheidungsfindung in Berlin durch "Filz" und Korruption unter besonderer Berücksichtigung der Bankgesellschaft Berlin AG*, Berlin: D.A. Freie Universität.

Cochrane, A. and Passmore, A. (2001) 'Building a National Capital in an Age of Globalisation', *Area*, vol. 33, no. 4, pp. 341-352.

Colás, A. (2005) 'Neoliberalism, Globalisation and International Relations', in Saad-Filho, A. and Johnston, J. (eds.) *Neoliberalism: A Critical Reader*, London: Pluto Press.

Collins, H. and Yearley, S. (1992) 'Epistemological Chicken', in Pickering, A. (ed.) *Science as Practice and Culture*, Chicago: University of Chicago Press.

Cook, I. R. (2009) 'Private sector involvement in urban governance: The case of Business Improvement Districts and Town Centre Management partnerships in England', *Geoforum*, vol. 40, no. 5, September, pp. 930-940.

Costabile-Heming, C., Halverson, R. and Foell, K. (2004) (eds.) *Berlin: The Symphony Continues: Orchestrating Architectural, Social and Artistic Change in Germany's New Capital*, Berlin: de Gruyter.

Cox, R. (1997) 'Democracy in hard times: economic globalisation and the limits to liberal democracy', in McGrew, A. (ed.) *The transformation of democracy? Globalization and Territorial Democracy*, Cambridge: Polity Press.

Crouch, C. (2004) *Post-democracy*, Cambridge: Polity Press.

Czarniawska, B. and Sevon, G. (2005) 'Translation is a Vehicle, Imitation its Motor, and Fashion sits at the Wheel', in Czarniawska, B. and Sevon, G. (eds.) *Global Ideas: How Ideas, Objects and Practices Travel in the Global Economy,* Copenhagen: Copenhagen Business School Press.

Danziger, M. (1995) 'Policy analysis postmodernized: Some political and pedagogical ramifications, *Policy Studies Journal*, vol. 23, pp. 435-450.

Dean, M. (2007) *Governing Societies: Political Perspectives on Domestic and International Rule*, McGraw Hill: Open University Press.

Dean, M. (2008) *Governmentality: power and rule in modern societies*, London: Sage.

Der Tagesspiegel (1997a) 'Berliner Wasserbetriebe vor dem Verkauf', 20 May.

Der Tagesspiegel (1997b) 'Verkauf der Wasserbetriebe gegen Berliner Verfassung', 11 September.

Der Tagesspiegel (1997c) 'Staatsbetriebe drängen ins Ausland', 16 September.

Der Tagesspiegel (1998) 'Wasserbetriebe sind begehrt', *Der Tagesspiegel*, 18 June.

Der Tagesspiegel (1999a) 'Verkauf der Wasserbetriebe schürt Angst vor Rationalisierung', 23 January.
Der Tagesspiegel (1999b) 'CDU stellt Verkauf der Wasserbetriebe in Frage', 18 March.

Der Tagesspiegel (1999b) 'Verkauf der Wasserbetriebe kurz vor dem Abschluss', 19 May.

Der Tagesspiegel (1999c) 'Stichwort', 11 August.

Der Tagesspiegel (1999d) 'Berlin fördert die Informationsgesellschaft', 21 January.

Doron, G. (1992) `Rational choice and policy sciences', *Policy Studies Review*, vol. 11, no.3, pp. 359-369.

Dryzek, D. (1994) *Discursive Democracy: politics, policy and political science*, Cambridge: Cambridge University Press.

Duménil, G. and Lévy, D. (2005) 'The Neoliberal (Counter)-revolution' in Saad-Filho, A. and Johnston, J. (eds.) *Neoliberalism: A Critical Reader*, London: Pluto Press.

Eckardt, F. (2004) 'Berlin: a Capital City between Nation State and Global Gateways', *City Futures,* Conference, University of Illinois, Chicago 8-10 July 2004, (Online), Available: http://www.uic.edu/cuppa/cityfutures/papers/webpapers/cityfuturespapers/session4_2/4_2berlin.pdf (21 November 2009).

Edie (2009) *Consultants expect public sector contracts to outstrip private sector,* (Online), Available: http://www.edie.net/library/view_article.asp?id=5227 (2 December 2009).

Edwards, P. N. (2001) 'Representing the Global Atmosphere: Computer Models, Data and Knowledge about Climate Change', in Miller, C. A. and Edwards, P. N. (eds.) *Changing the Atmosphere: Expert Knowledge and Environmental Governance,* Cambridge, MA: MIT Press.

Edwards, R. (2009) 'World Water Day', *The Herald,* 22 March, (Online), Available: http://www.heraldscotland.com/world-water-day-1.826734 (12 November 2009).

Ellger, F. (1992) 'Berlin: legacies of division and problems of unification', *The Geographical Journal,* vol. 158, no. 1, pp. 40-46.

Esser, J. (1998) 'Privatization in Germany: Symbolism in the Social Market Economy', in Parker, D. (ed.) *Privatisation in the European Union: Theory and Policy Perspectives,* London: Routledge Press.

European Federation of Management Consultancies Associations (2006) *Survey of the European Management Consultancy Market, Report 2006-7,* (Online), Available: http://www.admin.bdu.de/docs/downloads/Studien/Feaco/feaco%20survey%202007%20-%202008.pdf (10 November 2009).

European Federation of Management Consultancies Associations (2007) *Survey of the European Management Consultancy Market, Report 2007-8,* (Online), Available: http://www.admin.bdu.de/docs/downloads/Studien/Feaco/feaco%20survey%202007%20-%202008.pdf (10 November 2009).

Fine, B. (2005) 'From Actor Network Theory to Political Economy', *Capitalism Nature Socialism,* vol. 16, no. 4, pp. 91-108.

Fischer, F. (1998) 'Beyond Empiricism: Policy Inquiry in Post positivist Perspective', *Policy Studies Journal,* vol. 26, no. 1, pp. 129-46.

Fischer, F. (2003) *Reframing Public Policy: Discursive Politics and Deliberative Practices,* Oxford: Oxford University Press.

Fischer, F. (2009) *Democracy and Expertise: Reorienting Policy Inquiry,* Oxford: Oxford University Press.

Fischer, R. (1997) 'Berliner Wasserbetriebe werden AG', *Die Welt,* 14 August.

Fischer, R. (1999) 'Stühlerrücken bei Wasser-Betrieben', *Die Welt,* 3 November.

Fitch, K. (2007a) 'Liquidating the public sector? Water privatization in France and Germany', *Dissertation available from ProQuest,* Paper AAI3260904, (Online), Available: http://repository.upenn.edu/dissertations/AAI3260904 (1 January 2007).

Fitch, K. (2007b) 'Water privatisation in France and Germany: the importance of local interest groups', *Local Government Studies*, vol. 33, no.1, pp. 589-605.

Flinders, M. and Buller, J. (2006) 'Depoliticisation: Principles, Tactics and Tools', *British Politics*, vol. 1, no. 3, November, pp. 293-318.

Flyvbjerg, B. (2006) 'Five Misunderstandings About Case-Study Research', *Qualitative Inquiry*, vol. 12, no. 2, pp. 219-245.

Foucault, M. (1991) 'Governmentality', in Burchill, G., Gordon, C. and Miller, P. (eds.) *The Foucault Effect: Studies in Governmentality*, Chicago: University of Chicago.

Foucault, M. (1991) *Discipline and Punish: The Birth of the Prison*, London: Penguin.

Foucault, M. (1998) *The Will to Knowledge: the history of sexuality volume 1*, London: Penguin.

Frankfurter Allgemeine Zeitung (1997a) 'Berlins Wasserbetriebe erschließen neue Geschäftsfelder', 18 April.

Frankfurter Allgemeine Zeitung (1997b) 'Berliner Wasserbetriebe: Eher Börsengang als Privatisierung', 21 May.

Frankfurter Allgemeine Zeitung (1999a) 'Vivendi wirbt um Berliner Wasserbetriebe', 13 January.

Frankfurter Allgemeine Zeitung (1999b) 'Wasserbetriebe für 2 Milliarden DM', 27 January.

Freeman, R. (2007) 'Policy moves: translation, policy and politics', Conference Paper, *2nd International Conference in Interpretive Policy Analysis: Research and Practice*, 31 May - 2 June 2007, Amsterdam: the Netherlands.

Frese, A. (1999) 'Berlin braucht viel Geduld', *Der Tagesspiegel*, 13 February.

Fukuyama, F. (2006) 'The End of History', in Tuathail, G. O., Dalby, S. and Routledge, P. (eds.) *The Geopolitics Reader*, London: Routledge Press.

Füller, C. (1997) 'Umkämpfte Wasserbetriebe', *Die Tageszeitung*, 13 August.

Gandy, M. (1997) 'The making of a regulatory crisis: the restructuring of New York City's water supply', *Transactions of the Institute of British Geographers*, vol. 22, no. 3, pp. 338-358.

Gandy, M. (2004) 'Rethinking urban metabolism: Water, space and the modern city', *City*, vol. 8, no. 3, December, pp. 363-379.

Gandy, M. (2005) 'Cyborg urbanization: complexity and monstrosity in the contemporary city', *International Journal of Urban and Regional Research*, vol. 29, no. 1, March, pp. 26-49.

Garrety, K. (1997) 'Social Worlds, Actor-Networks and Controversy: The Case of Cholesterol, Dietary Fat and Heart Disease', *Social Studies of Science*, vol. 27, no. 5, pp. 727-773.

Gleeson, B. and Low, N. (2000) 'Cities as consumers of the world's environment', in Low, N., Gleeson, B., Elander, I. and Lidskog, R. (eds.) *Consuming Cities*, London: Routledge Press.

Global Water Intelligence (2008) *Lobby group calls for Scottish Water privatisation*, vol. 9, no. 11, (Online), Available: http://www.globalwaterintel.com/archive/9/11/general/lobby-group-calls-for-scottish-water-privatisation.html (30 September 2009).

Glynos, J. and Howarth, D. (2008) 'Structure, Agency and Power in Political Analysis: Beyond Contextualised Self-Interpretations', *Political Studies Review*, vol. 6, no. 2, pp. 155-169.

Goldblatt, D., Held, D., McGrew, A. G. and Perraton, J. (1997) 'Economic Globalisation and the Nation-State: Shifting Balances of Power', *Soundings*, vol. 7, pp. 61-77.

Goldman, M. (2007) 'How "Water for All!" policy became hegemonic: The power of the World Bank and its transnational policy networks', *Geoforum*, vol. 38, no. 5, pp. 786-800.

Gomart, E. and Hajer, M. (2003) 'Is That Politics? For an Inquiry Into Forms in Contemporary Politics', in Joerges, B. and Nowotny, H. (eds.) *Social Studies of Science and Technology: Looking Back, Ahead*, Dordrecht/Boston: Kluwer Academic Publications.

Gordon, C. (1991) 'Governmental Rationality: an Introduction' in Burchill, G., Gordon, C. and Miller, P. (eds.) *The Foucault Effect: Studies in Governmentality*, Chicago: University of Chicago.

Gornig, M. and Haeussermann, H. (2002) 'Berlin: Economic and Social Change', *European Urban and Regional Studies*, vol. 9, no. 4, pp. 331-341.

Gottweis, H. (1999) 'Regulating genetic engineering in the European Union: a post-structuralist perspective', in Kohler-Koch, B. and Eising, R. (eds.) *The Transformation of Governance in the European Union*, London: Routledge Press.

Gottweis, H. (2003) 'Theoretical Strategies of Post-Structuralist Policy Analysis: Towards an Analytics of Government,' in Hajer, M. and Wagenaar, H. (eds.) *Deliberative Policy Analysis: Understanding Governance in the Network Society*, Cambridge: Cambridge University Press.

Gow, D. (2009) 'German Rail Privatisation runs out of steam', *The Guardian*, 2 April, (Online), Available: http://www.guardian.co.uk/business/2009/apr/02/europe-deutschebank (5 January 2009).

Griffiths, K. (2006) 'Fury at Brown Plans for Scottish Water', *The Daily Telegraph* 21 March, (Online), Available: http://www.telegraph.co.uk/finance/2934740/Fury-at-Brown-plans-for-Scottish-Water.html (21 October 2009).

Grin, J. and Loeber, A. (2007) 'Theories of Policy Learning: structure, agency and change' in Fischer, F., Miller, G. J. and Sidney, M. S. (eds.) *Handbook of public policy analysis: theory, politics and methods*, London: Taylor and Francis.

Grindle, M. S. (1977) 'Power, Expertise and the "Tecnico": Suggestions from a Mexican Case Study', *The Journal of Politics* (Electronic), vol. 39, no. 2, pp. 399-426, (Online), Available: http://www.jstor.org/stable/2130057, (16 March 2009).

Guttman, D. and Willner, B. (1976) *The Shadow Government: The Government's Multi-Billion-Dollar Giveaway of Its Decision-Making Powers to Private Management Consultants, 'Experts', and Think Tanks*, New York: Pantheon Books.

Hacking, I. (1999a) *The Social Construction of what?*, London: Harvard University Press.

Hacking, I. (1999b) *Mad Travellers: reflections on the reality of transient mental illness*, London: Free Association Books.

Häussermann, H. and Colomb, C. (2003) `The New Berlin: Marketing the City of Dreams', in Hoffman, L. M., Fainstein, S. S. and Judd, D. R. (eds.) *Cities and Visitors: Regulating People, Markets, and City Space*, New York, NY: Blackwell Publishing.

Hall, D. (1999) *Privatisation, multinationals and corruption, Public Services International Research Unit (PSIRU) reports*, (Online), Available: http://www.psiru.org/reports/9909-U-U-Corrup.doc (12 August 2007).

Hall, D. (2001) 'Water Privatisation: global problems, global resistance', *Public Services International Research Unit (PSIRU)*, (Online) Available: http://www.psiru.org/reports/2001-07-W-salb.doc (1 December 2009).

Hall, D., Lobina, E. and de la Motte, R. (2004) 'Making privatisation illegal: New laws in the Netherlands and Uruguay', *Public Services International Research Unit (PSIRU) report,* (Online) Available: www.psiru.org/reports/2004-11-W-crim.doc. (22 February 2010).

Hall, D., Lobina, E. and de la Motte, R. (2005) 'Public resistance to privatisation in water and energy', *Development in Practice*, vol. 15, no. 3-4, pp. 286-301.

Hall, D. and Lobina, E. (2005) 'The relative efficiency of public and private water', *Public Services International Research Unit (PSIRU) report,* (Online), Available: http://www.psiru.org/reports/2005-10-W-effic.doc, (1 December 2009).

Hall, D. and Lobina, E. (2008) 'Water in Europe', *Public Services International Research Unit (PSIRU) reports*, (Online), Available: http://www.psiru.org/publicationsindex.asp (20 March 2010).

Hajer, M. (1997) *The Politics of Environmental Discourse*, Oxford: Oxford University Press.

Hajer, M. (2002) 'Discourse Analysis and the Study of Policy-making', *European Political Science,* vol. 2, no.1, pp. 61-65.

Hajer, M. (2003a) 'A Frame in the Fields: Policy-making and the Reinvention of Politics', in Hajer, M. and Wagenaar, H. (eds.) *Deliberative Policy Analysis. Understanding Governance in Network Society*, Cambridge: Cambridge University Press.

Hajer, M. (2003b) 'Policy without Polity? Policy analysis and the institutional void', *Policy Sciences*, vol. 36, no. 2, pp. 175-195.

Hajer, M. and Fischer, F. (1999) 'Beyond Global Discourse: the rediscovery of culture in environmental politics', in Fischer, F. & Hajer, M. (eds.) *Living with nature: environmental politics as cultural discourse*, Oxford: Oxford University Press.

Hajer, M. and Kesselring, S. (1999) 'Democracy in the Risk Society? Learning from the New Politics of Mobility in Munich', *Environmental Politics*, vol. 3, pp. 1-23.

Hajer, M. and Wagenaar, H. (eds.) (2003) *Deliberative Policy Analysis: Understanding Governance in the Network Society*, Cambridge: Cambridge University Press.

Hajer, M. and Versteeg, W. (2005) 'Performing Governance Through Networks', *European Political Science*, vol. 4, no. 3, pp. 340-347.

Hansen, A. and Sorensen, E. (2005) 'Polity as Politics: Studying the Shaping and Effects of Discursive Polities', in Howarth, D. and Torfing, J. (eds.) *Discourse Theory in European Politics: Identity, Policy and Governance*, London: Palgrave Macmillan.

Handelsblatt (1997) 'Franzosen zeigen Interesse', 21 May.

Harvey, D. (1989) 'From managerialism to entrepreneurialism: The transformation in urban governance in late capitalism', *Geografiska Annaler*, vol. 71 B. no.1, pp. 3-17.

Harvey, D. (2005) *A Brief History of Neo-liberalism*, Oxford: Oxford University Press.

Hassan, J. (1998) *A History of Water in Modern England and Wales*, Manchester: Manchester University Press.

Hay, C. (2007) *Why We Hate Politics*, Cambridge: Polity Press.

Hay, C. and Rosamond, B. (2002) 'Globalisation, European Integration and the Discursive Construction of Economic Imperatives', *Journal of European Public Policy*, vol. 9, no. 2, pp. 147-167.

Held, D. and McGrew, A. (2003) 'The Great Globalisation Debate: An Introduction', *The Global Transformations Reader: an introduction to the globalization debate*, Cambridge: Polity Press.

Held, D. (2006) *Models of Democracy*, Cambridge: Polity Press.

Heynen, N., Kaika, M. and Swyngedouw, E. (2005) (eds.) *In the Nature of Cities*, London: Routledge Press.

Heynen, N., McCarthy, J., Prudham, S. and Robbins, P. (2007) (eds.) *Neoliberal Environments: False Promises and Unnatural Consequences*, Abingdon, Oxon: Routledge Press.

Hill, M. (1997) 'Introduction: theories of the State', in Hill, M. (ed.) *The Policy Process: a reader*, Harlow, UK: Pearson Education Limited.

Hodge, G. and Bowman, D. (2006) 'The 'Consultocracy': the business of reforming government' in Hodge, G. (ed.) *Privatization and Market Development: global movements in public policy*, Cheltenham: Edward Elgar Publishing Ltd.

Hoffbauer, A. (1998) 'Verwässerter Kompromiss', *Berliner Morgenpost*, 7 July.

Howarth, D. (2005) 'Applying discourse theory: the method of articulation', in Howarth, D. and Torfing, J. (eds.) *Discourse Theory in European Politics: Identity, Policy and Governance*, London: Palgrave Macmillan.

Hüesker, F. (2011): Kommunale Daseinvorsorge in der Wasserwirtschaft: Auswirkungen der Privatisierungen am Beispiel der Wasserbetriebe Berlins. München: Oekom verlag.

Jasanoff, S. (2006) (ed.) 'Ordering Knowledge, Ordering Society', in Jasanoff, S. (ed.) *States of Knowledge: The Co-production of Science and Social Order*, London: Routledge Press.

Jessop, B. (2000) 'Globalization, entrepreneurial cities and the social economy', in Hamel, P., Lustiger-Thaler, H. and Mayer, M. (eds.) *Urban Movements* in a *Globalizing World*, London: Routledge Press.

John, P. (2003) 'Is there Life After Policy Streams, Advocacy Coalitions and Punctuations: Using Evolutionary Theory to explain Policy Change?' *The Policy Studies Journal*, vol. 32, no. 4, pp. 481-498.

Jordan, A., Wurzel R. and Zito, A. (2005) 'The rise of 'new' policy instruments in comparative perspective: has governance eclipsed government?', *Political Studies*, vol. 53, no. 3, pp. 477-496.

Kaika, M. (2004) *City of Flows: modernity, nature and the city*, London: Routledge Press.

Kane, T. and Russell, S. (2007) 'Is anything public anymore?' *Scottish Left Review*, vol. 40, (Online), Available: http://www.scottishleftreview.org/li/index.php?option=com_content&task=view&id=31&Itemid=29 (20 November 2009).

Kay, J. A. and Thompson, D. J. (1986) 'Privatisation: a policy in search of a rationale' *The Economic Journal*, vol. 96, pp. 18-32.

Kendall, G. (2004) 'Global networks, international networks, actor networks', in Larner, W. and Walters, W. (eds.) *Global governmentality: governing international spaces*, London: Routledge Press.

King, G., Koehane, R. and Verba, S (1994) Designing Social Inquiry: Scientific Inference in Qualitative Research, Princeton: Princeton University Press.

Klingemann, H-D and Hofferbert, R. (1994) *Parties, Policies, And Democracy (Theoretical Lenses on Public Policy)*, Boulder, Col.: Westview Press.

Kooiman, J. (2003) *Governing as governance*, London: Sage.

Kraetke, S. (2001) 'Berlin - Towards a Global City?', *Urban Studies*, vol. 38, no. 10, pp. 1777-1799.

Kraetke, S. (2004a) 'City of Talents? Berlin's regional economy, socio-spatial fabric and 'worst practice' urban governance', *International Journal of Urban and Regional Research*, vol. 28, no. 3, pp. 511-529.

Kraetke, S. (2004b) 'Economic restructuring and the making of a financial crisis: Berlin's socio-economic development path 1989 - 2004', *DISP - Special Issue on "Berlin 15 Years after the breaking of the wall"*, Zürich.

Lanz, K. and Eitner, K. (2005a) *Watertime case study: Berlin, Germany*, Available: http://www.watertime.net/docs/WP2/D12_Berlin.doc (24August 2007).

Lanz, K. and Eitner, K. (2005b) 'Background excel document to Berlin Case Study', *Watertime* Project, unpublished.

Larner, W. (2000) 'Neo-liberalism: Policy, Ideology, Governmentality', *Studies in Political Economy*, vol. 63, pp. 5-26.

Larner, W. and Le Heron, R. (2004) 'Global benchmarking: participating 'at a distance' in the global economy', in Larner, W. and Walters, W. (eds.) *Global governmentality: governing international spaces*, London: Routledge Press.

Larner, W. and Walters, W. (2004) 'Introduction: global governmentality', in Larner, W. and Walters, W. (eds.) *Global governmentality: governing international spaces*, London: Routledge Press.

Latham, A. (2006) 'Anglophone Urban Studies and the European City: some comments on interpreting Berlin', *European Urban and Regional Studies*, vol. 13, no. 11, pp. 88-92.

Latour, B. (1987) *Science in Action: How to Follow Scientists and Engineers Through Society*, Milton Keynes: Open University Press.

Latour, B. (1993) *We Have Never Been Modern*, Cambridge, MA: Harvard University Press.

Latour, B. (1999) 'On Recalling ANT', in Law, J. and Hassard, J. (eds.) *Actor Network and After*, Oxford: Blackwell and the Sociological Review.

Latour, B. (2005) *Reassembling the social: An Introduction to Actor-Network Theory*, Oxford: Oxford University Press.

Law, J. (1991) 'Power, Discretion and Strategy', in Law, J. (ed.) *A Sociology of Monsters; Essays on Power, Technology and Domination, Sociological Review Monograph 38*, London: Routledge Press.

Law, J. (1992) 'Notes on the theory of the Actor-network: ordering, strategy and heterogeneity', *Systems Practice*, vol. 5, no. 4, pp. 379-393.

Law, J. (1994) *Organizing Modernity: Social Action and Social Theory*, Oxford: Blackwell.

Law, J. (1999) 'After ANT: complexity, naming and topology', in Law, J. and Hassard, J. (eds.) *Actor Network Theory and After*, Oxford: Blackwells.

Law, J. and Singleton, V. (2000) 'Performing Technology's Stories', published by the *Centre for Science Studies, Lancaster University*, Lancaster LA1 4YN, UK, (Online), Available: http://www.comp.lancs.ac.uk/sociology/papers/Law-Singleton-Performing-Technology's-Stories.pdf (22 March 2010).

Lemke, T. (2001) 'The Birth of Bio-Politics: Michel Foucault's Lectures at the College de France on Neo-Liberal Governmentality', Author's personal website, (Online), Available: http://www.thomaslemkeweb.de/engl.%20texte/The%20Birth%20of%20Biopolitics%203.pdf (December 2008).

Letza, S. R., Smallman, C. and Sun, X. (2004) 'Reframing privatisation: deconstructing the myth of efficiency', *Policy Sciences* vol. 37, pp. 159-183.

Leitner, H., Sheppard, E. S., Sziarto, K. and Maringanti, A. (2007) 'Contesting urban futures: de-centring neoliberalism', in Leitner, H., Peck, J. and Sheppard, E. S. (eds.) *Contesting Neoliberalism: Urban Frontiers*, New York: Guilford.

Lijphart, A (1999) *Patterns of Democracy: Government Forms & Performance in Thirty-six Countries*, New Haven: Yale University Press.

Lobina, E. (2001) 'Water privatisation and restructuring in Central and Eastern Europe', *Public Services International Research Unit (PSIRU) report,* (Online), Available: http://www.psiru.org/publicationsindex.asp (23 March 2010).

Lynn, L. (1999) *A Place at the Table: Policy Analysis, Its Postpositive Critics, and the Future of Practice*, Draft paper. Available: http://harrisschool.uchicago.edu/about/publications/working-papers/pdf/wp_99_01.pdf (20 June 2010).

MacRae, D. and Whittington, D. (1997) *Expert Advice for Policy Choice: Analysis and Discourse*, Washington, DC: Georgetown University Press.

Maloney, W. (2001) 'Regulation in an Episodic Policy-Making Environment: The Water Industry in England and Wales', *Public Administration*, vol. 79, no. 3, pp. 625-642.

Maloney, W. and Richardson, J. J. (1995) 'Water Policy-making in England and Wales: policy communities under pressure', in Bressers, H., O'Toole, L. and Richardson, J. (eds.) *Networks for water policy: a comparative perspective*, London: Frank Cass.

Marcuse, P. (1998) 'Reflections on Berlin: the meaning of construction and the construction of meaning', *International Journal of Urban and Regional Research*, vol. 22, no. 2, pp. 331-338.

Marsh, D. and Rhodes, R. A. W. (eds.) (1992) *Policy Networks in British Government*, Oxford: Clarendon Press.

Marschall, M., Richter, C. and Keese, C. (1997) 'Nach Ostern muss Entscheidung über die Wasserbetriebe fallen', *Berliner Zeitung*, 17 March.

Martin D., McCann E. and Purcell, M. (2003) 'Space, scale, governance and representation: contemporary geographical perspectives on urban politics and policy', *Journal of Urban Affairs*, vol. 25, no. 2, pp. 113-121.

Mayntz, R. (2003) 'New challenges to governance theory', in Bang, H. (ed.) *Governance as political and social communication*, Manchester: Manchester University Press.

McCann, E. J. (2001) 'Collaborative visioning or urban planning as therapy? The politics of public-private policy-making', *Professional Geographer*, vol. 53, no. 2, pp. 207-18.

McKee, K. (2009) 'Post-Foucauldian governmentality: what does it offer critical social policy analysis?', *Critical Social Policy*, vol. 29, no. 3, pp. 465-486.

Mclean, C. and Hassard, J. (2004) 'Symmetrical Absence/ Symmetrical Absurdity: Critical notes on the production of Actor- Network Accounts' *Journal of Management Studies*, vol. 41, no. 3, pp. 494-515.

McNabb, D. (2004) *Research Methods for Political Science: Quantitative and Qualitative Methods*, Armonk, N.Y.: M.E. Sharpe.

Megginson, W. L. (2005) *The Financial Economics of Privatization*, Oxford/ New York: Oxford University Press.

Megginson, W. L., Nash, R. C. and van Randenborg, M. (1996) 'The Financial and Operating Performance of Newly Privatised Firms: An International Empirical Analysis', in Anderson, T. L. and Hill, P. J. (eds.) *The Privatisation process: A Worldwide Perspective*, Lanham, MD: Rowan Littlefield.

Megginson, W. L. and Netter, J. F. (2001) 'From state to market: A survey of empirical studies on privatization', *Journal of Economic Literature*, vol. 39, pp. 321–389.

Miller, C. A. and Edwards, P. N. (2001) (eds.) *Changing the Atmosphere: Expert Knowledge and Environmental Governance,* Cambridge, MA: MIT Press.

Miller, H. T. and Fox, C. J. (2007) *Postmodern Public Administration*, New York: M. E. Sharpe.

Miller, P. and Rose, N. (2008) *Governing the Present*, Cambridge: Polity Press.

Mohajeri, S. (2006) 'Die Privatisierung der Berliner Wasserbetriebe damals and heute: Eine kritische Betrachtung', in Frank, S. and Gandy, M. (eds.) *Hydropolis: Wasser und die Stadt der Moderne*, Frankfurt/ Main: Campus.

Monstadt, J. (2007) 'Urban governance and the transition of energy systems: institutional change and shifting energy and climate policies in Berlin', *International Journal for Urban and Regional Research*, vol. 31, no. 2, pp. 326-343.

Moran, M. (2006) 'Economic Institutions', in Rhodes, R., Binder, S. and Rockman, B., (eds.) *Oxford Handbook of Political Institutions*, Oxford: Oxford University Press.

Moss, T. (2009) 'Divided City, Divided Infrastructures: Securing Energy and Water Services in Postwar Berlin', *Journal of Urban History*, vol. 35, no.7, pp. 923-942.

Mouffe, C. (2005) *On the Political*, Abingdon, Oxon: Routledge Press.

Munck, R. (2005) 'Neoliberalism and Politics, and the Politics of Neoliberalism', in Saad-Filho, A. and Johnston, J. (eds.) *Neoliberalism: A Critical Reader*, London: Pluto Press.

Murdoch, J. (2001) 'Ecologising Sociology: Actor-Network Theory, Co-Construction and the Problem of Human Exemptionalism', *Sociology*, vol. 35, no. 1, pp. 111-133.

Murdoch, J. (2003) 'Co-constructing the countryside: hybrid networks and the extensive self', in Cloke, P. (ed.) *Country Visions,* London: Pearson Education.

Newman, P. and Thornley, A. (1996) *Urban Planning in Europe: international competition, national systems and planning projects*, London: Taylor and Francis.

Naumann, M. (2009) *Neue Disparitäten durch Infrstruktur? Der Wandel der Wasserwirtschaft in ländlich-peripheren Räumen*, München: oekom.

O'Malley, P., Weir, L. and Shearing, C. (1997) 'Governmentality, Criticism, Politics', *Economy and Society*, vol. 26, no. 4, pp. 501–17.

Ostrom, E., (1997) 'A behavioral approach to the rational choice theory of collective action: Presidential address, American Political Science Association', *American Political Science Review*, vol. 92, pp. 1-22.

Passadakis, A. (2006), 'The Berlin Water Works: from commercialisation and partial privatisation to public democratic water enterprise', *Fraktion der Vereinigten europäischen Linken/Nordische Grüne Linke im Europäischen Parlament*, Brussels.

Peters, B. G. and Savoie, D. J. (1998) 'Introduction', in Peters, B. G. and Savoie, D. J. (eds.) *Taking Stock: Assessing Public Sector Reforms*, Montreal and Kingston: McGill-Queen's University Press.

Peck, J. and Tickell, A. (2002) 'Neoliberalizing Space', in Brenner, N. and Theodore, N. (eds.) *Spaces of Neoliberalism: urban restructuring in North America and Western Europe*, Oxford: Blackwell.

Peck, J. (2004) 'Geography and public policy: constructions of neoliberalism', *Progress in Human Geography*, vol. 28, no. 3, pp. 392-405.

Poole, R. W. (1996) 'Privatisation for economic development', in Anderson, T. L. and Hill, P. J. (eds.) *The Privatisation process: A Worldwide Perspective*, Lanham, MD: Rowan Littlefield.

Public Citizen (2005) 'Veolia Environment: a corporate profile', *Public Citizen* report, (Online), Available: www.citizen.org/documents/vivendi-uSFilter.pdf (2 December 2009).

Putt, A. and Springer, F. (1989) *Policy Research: Concepts, Methods, and Applications*, New Jersey: Prentice-Hall.

Rawnsley, A. (2010) 'Despite their hopes for a great revival, the left got left behind', *Guardian* (13 June 2010).

Reitemeier, D. (1999) 'Verkauf der Wasserbetriebe abgehakt', *Berliner Morgenpost*, 19 June.

Reuters (2009) *Italy Parliament approves water privatisation law,* (Online), Available: http://uk.reuters.com/article/idUKLJ39968220091119 (3 December 2009).

Richardson, J. (2000) 'Government, Interest Groups and Policy Change', *Political Studies*, vol. 48, pp. 1006-1025.

Riecker, J. (1997) 'Geheimpapier der Berater-Firma: Berliner Wasser ist zu teuer', *Berliner Morgenpost*, 26 August.

Riecker, J. (1998) 'Zukunft der Wasser-Betriebe ungeklärt', *Berliner Morgenpost*, 14 January.

Riecker, J. and Scharf, R. (1998) 'SPD bremst Privatisierung', *Berliner Morgenpost*, 4 June.

Riecker, J. (1999a) 'Kampf um Teures Wässerchen', *Berliner Morgenpost*, 4 February.

Riecker, J. (1999b) 'Streit um BWB-Verkauf hält an', *Berliner Morgenpost*, 18 March.

Riecker, J. (1999c) 'Mieter und Vermieter warnen vor BWB Verkauf', *Berliner Morgenpost*, 20 March.

Riecker, J. (1999d) 'Geheimangebot: Jobgarantie für 15 Jahre', *Berliner Morgenpost*, 17 April.

Riecker, J. (1999e) 'OTV will Job-Garantie oder Streik', *Berliner Morgenpost,* 20 April.

Rhodes, R. A. W. (1990) 'Policy Networks: a British perspective', *Journal of Theoretical Politics*, vol. 2, no. 3, pp. 292-316.

Rhodes, R. A. W. (2006) 'Policy Network Analysis', in Moran, M., Rein, M. and Goodin, R. E. (eds.) *The Oxford Handbook of Public Policy*, Oxford: Oxford University Press.

Rose, M. D. (2004) *Warten auf die Sintflut*, Berlin: Transit.

Rose, N. and Miller, P. (1992) 'Political Power Beyond the State: Problematics of Government', *British Journal of Sociology*, vol. 43, no. 2, June, pp. 173-205.

Rose-Ackerman, S. (1999) *Corruption and Government: causes, consequences and reform*, Cambridge: Cambridge University Press.

Rutland, T. and Aylett, A. (2008) 'The work of policy: actor networks, governmentality and local action on climate change', *Environment and Planning D: Society and Space*, vol. 26, pp. 627-646.

Sabatier, P. (ed.) (1999) *Theories of the Policy Process*, Boulder, Co: Westview Press.

Sabatier, P. (1999) 'The Need for Better Theories', in P. Sabatier (ed.) *Theories of the Policy Process*, Boulder, Co: Westview Press.

Sabatier, P. (2000) 'Clear Enough to Be Wrong', *Journal of European Public Policy*, vol. 7, pp. 137-142.

Sabatier, P. and Jenkins-Smith, H. (1993) 'The Advocacy Coalition Framework: assessment, revisions and implications for scholars and practitioners', in Sabatier, P. and Jenkins-Smith, H. (eds.) *Policy Change and Learning: An Advocacy Coalition Approach*, Boulder, Co: Westview Press.

Sabatier, P. and Jenkins-Smith, H. (1999) 'The Advocacy Coalition Framework: An Assessment', in Sabatier, P. (ed.), *Theories of the Policy Process*, Boulder, Co: Westview Press.

Saad-Filho, A. and Johnston, J. (2005) (eds.) *Neoliberalism: A Critical Reader*, London: Pluto Press.

Saint-Martin, D. (2000) *Building the New Managerialist State*, Oxford: Oxford University Press.

Sassen, S. (2001) *The Global City*, Princeton/ Woodstock, Oxfordshire: Princeton University Press.

Sauri, D., Olcina, J. and Rico, A. (2007) 'The March towards privatisation: Urban Water Supply and Sanitation in Spain', *Journal of Comparative Social Welfare*, vol. 23, no. 2, pp. 131-139.

Scharf, R. (1999) 'SPD-Vize attackiert Fugmann-Heesing', *Berliner Morgenpost*, 20 March.

Schedler, K. and Proeller, I. (2002) 'The New Public Management: a perspective from mainland Europe', in McLaughlin, K., Osborne, S. P. and Ferlie, E. (eds.) *New Public Management: current trends and future prospects*, London: Routledge Press.

Schmitter, P. (1974) 'Still the Century of Corporatism?' *The Review of Politics*, vol. 36, no. 1, pp. 85-131.

Schomaker, G. (1998) 'Beraten oder Verraten', *B.Z.*, 20 June.

Schomaker, G. (1999) 'Rechnungshof-Präsident warnt vor neuen Schulden', *Berliner Zeitung*, 28 October.

Schramm, E. (2004) 'Privatisation of German urban water infrastructure in the 19th and 21st century', in Wilding, P. (ed.) *Urban Infrastructure in Transition: What can we learn from history? International Summer Academy on Technology Studies*, 11-17 July 2004, Deutschlandsberg, Austria, Graz (A): IFF/IFZ, (Online), Available: http://www.isoe.de/ftp/ESDlandsberg.pdf (4 January 2009).

Schulte, E. (1999a) 'Wasserbetriebe: Land will nur einen Bruchteil des Erlöses versteuern', *Berliner Zeitung*, 15 February.

Schulte, E. (1999b) 'Wasserbetriebe: "Maulkorb" für die Investoren', *Berliner Zeitung*, 18 February.

Schulte, E. (1999c) 'Preis für Wasser soll in Berlin stabil bleiben', *Berliner Zeitung*, 20 March.

Schulte, E. (1999d) 'Der milliardenschwere Poker um die Wasserbetriebe', *Berliner Zeitung*, 31 May.

Schulte, E. (1999e) 'Wasserbetriebe: Diepgen setze RWE –Votum durch', *Berliner Zeitung*, 9 June.

Shaoul, J., Stafford, A. and Stapleton P. (2007) 'Private control over public policy: financial advisors and the private finance initiative', *Policy and Politics* vol. 35, pp. 479-496.

Shore, C. and Wright, S. (1997) 'A new field of anthropology', in Shore, C. and Wright, S. (eds.) *Anthropology of Policy: Critical Perspectives on Governance and Power*, Abingdon, Oxon: Routledge Press.

Spannbauer, A. (1999) 'Berliner Wasserbetriebe: Alles unklar', *Die Tageszeitung*, 6 July.

Streeck, W. (2006) 'The study of organised interests: before "The Century" and after', in Crouch, C. and Streeck, W. (eds.) *The Diversity of Democracy: Corporatism, Social Order and Political Conflict*, London: Edward Elgar.

Stoker, G. (1998a) 'Governance as theory: five propositions', *International Social Science Journal*, no. 155, April, pp. 17-28.

Stoker, G. (1998b) 'Public-private partnerships and urban governance', in Pierre, J. (ed.) *Partnerships in Urban Governance: European and American Experience*. London, UK: Macmillan.

Strom, E. (2001) *Building the New Berlin: the Politics of Urban Development in Germany's Capital City*, Lanham, MD: Lexington Books.

Swyngedouw, E. (2003) *Privatising H2O Turning Local Water Into Global Money*, (Online), Available: http://socgeo.ruhosting.nl/colloquium/water.pdf (20 March 2010).

Swyngedouw, E. (2004) 'Globalisation or "Glocalisation"? Networks, Territories and Rescaling', *Cambridge Review of International Affairs*, vol. 17, no. 1, pp. 25-48.

Swyngedouw, E. (2005) 'Governance Innovation and the Citizen: The Janus Face of Governance-beyond-the-State', *Urban Studies*, vol. 42, no. 11, pp. 1991-2006.

Swyngedouw, E. (2006) 'Circulations and Metabolisms: (Hybrid) Natures AND (Cyborg) Cities', *Science as Culture*, vol. 15, no. 2, pp. 105-122.

Swyngedouw, E. (2007a) *Where is the political?*, Working paper based on Antipode Lecture.

Swyngedouw, E. (2007b) *Privatising H2O: Turning Local Waters into Global Money*,(Online), Available: http://socgeo.ruhosting.nl/colloquium/water.pdf (10 December 2009).

Swyngedouw, E. (2009a) 'The Antinomies of the Post-Political City: In search of a democratic politics of environmental production', *International Journal of Urban and Regional Research*, vol. 33, no. 3, September, pp. 601-620.

Swyngedouw, E. (2009b) 'Troubled Waters. The Political Economy of Essential Public Services', in Castro, J. E. and Heller, L. (eds.) *Water and Sanitation services: public policy and management*, London: Earthscan.

Swyngedouw, E. (2010) 'Apocalypse Forever? Post-political Populism and the Spectre of Climate Change', *Theory, Culture & Society*, vol. 27, no. 2–3, pp. 213–232.

Swyngedouw, E., Heynen, N. and Kaika (2006) (eds.) *In the Nature of Cities: Urban Political Ecology and the Politics of Urban Metabolism*, London: Routledge Press.

The Federal Association of German Consultants, (BDU), (Online), Available: http://www.bdu.de/Uebersicht_FV.html?fuseaction=page.content&s_kurzname=FV_OeA (10 November 2009).

Torfing, J. (2005) 'Discourse Theory: Achievements, Arguments and Challenges', in Howarth, D. and Torfing, J. (eds.) *Discourse Theory in European Politics: Identity, Policy and Governance*, London: Palgrave Macmillan.

Thrift, N. (1999) 'Cities and economic change: global governance?', in Allen, J., Massey, D. and Pryke, M. (eds.) *Unsettling cities: movement/ settlement*, London: Routledge Press.

Verheyen, D. (1999) *The German Question: A Cultural, Historical And Geopolitical Exploration*, Boulder/Colorado: Westview Press.

Vogelsang, I. (1988) 'Deregulation and Privatization in Germany', *Journal of Public Policy*, vol. 8, no. 2, pp. 195-212.

Von Törne, L. (2009) 'Senat: Externe Gutachter sparen mehr als sie kosten, Der Tagesspiegel, 19 August.

Von Weizsaecker, E. (2005) 'Post-war history: the ups-and-downs of the public sector', in Von Weizsaecker, E. U., Young, O. and Finger, M. (eds.) *Limits to privatization: how to avoid too much of good thing*, London: Earthscan.

Voss, K. (1999) 'Wasserbetriebe-Verkauf verzögert sich', *Der Tagesspiegel*, 13 March.

Walters, W. (2004) 'The Political Rationality of European Integration', in Larner, W. and Walters, W., (eds.) *Global governmentality: governing international spaces*, London: Routledge Press.

Ward, J. (2004) 'Berlin, the Virtual Global City', *Journal of Visual Culture*, vol. 3, no. 2, pp. 239-256.

Watson, M. and Hay, C. (2003) 'The Discourse of Globalisation and the Logic of No Alternative: Rendering the Contingent Necessary in the Political Economy of New Labour', *Policy and Politics*, vol. 30, no. 4, pp. 289-305.

Wedel, J., Shore, C., Feldman, G. and Lathrop, S. (2005) 'Toward an Anthropology of Public Policy', *Annals of the American Academy of Political and Social Science*, vol. 600, The Use and Usefulness of the Social Sciences: Achievements, Disappointments and Promise, July, pp. 30-51.

Weimer, David (1999) 'Comment: Q-Method and the Isms', *Journal of Policy Analysis and Management* vol. 18, pp. 426–9.

Werle, H. (2004) 'Between Public well-being and Profit interests', *Brot fuer* Die Welt *report*, (Online), Available: http://www.menschen-recht-wasser.de/downloads/Background_paper__Water_privatisation_in_Berlin.pdf (24 November 2009).

Williamson, J. (2000) 'What should the world bank think about the Washington consensus?', *The World Bank Research Observer*, vol. 15, no. 2, August, pp. 251-264.

Wetzel, D. (1998) 'SPD will Wasserbetriebe nur häppchenweise verkaufen', *Der Tagesspiegel*, 21 June.

Wiengten, K. (1997) 'Kein Wasser auf den Hießen Stein', *Die Tageszeitung*, 19 December.

Wilson, C. (2000) 'Policy Regimes and Policy Change', *Journal of Public Policy*, vol. 20, no. 3, pp. 247-274.

Wilson, J. (1986) *Banking Policy and Structure: A Comparative Analysis*, New York: New York University Press.

Wiskow, V.J-H. (1999) 'Der Skandal um das Berliner Wasser, *Der Tagesspiegel*, 23 March.

Wissen, M. and Naumann, M. (2006) *A new logic of infrastructure supply: the commercialization of water and the transformation of urban governance in Germany*, Draft of paper later published in *Social Justice* (2006), vol. 33, no. 3 (provided by authors).

Wuschick, D. (1997) 'Streit um Wasserwerke wird Chefsache', *Die Welt*, 17 July.

Yanow, D. (1995) 'Practices of policy interpretation', *Policy Sciences*, vol. 28, pp. 111-126.

Yanow, D. (2003) *How does a Policy mean? Interpreting Policy and Organizational Actions*, Georgetown: Georgetown University Press.

Zeiger, A. (1999) 'Wasserbetriebe: Einigung nicht in Sicht, *Die Welt*, 23 February.

Appendix: list of interviewees

Interview 1: ATTAC Berlin campaigner, 15 February 2006.

Interview 2: senior politician in the Berlin Left Party (*Die Linke*), 27 April 2006.

Interview 3: member of the BWB Management Board during the privatisation process, 31 March 2006.

Interview 4: Berlin-based journalist, 10 May 2006, Berlin.

Interview 5: civil servant at the Senate for City Development, Construction, Environment and Transport, who worked on the environmental aspects of the privatisation, 28 June 2006.

Interview 6: politician in the Green Party, *Die Grünen*, 14 August 2006.

Interview 7: senior Left Party Politician (also a member of parliament in the 1990s), *Die Linke*, 14 August 2006.

Interview 8: leading consultant for the RWE/Vivendi/Allianz consortium, 18 September 2006.

Interview 9: consultant hired by Finance Senate for the BWB bidding process, 4 August 2006.

Interview 10: BWB workers' representative during the privatisation, 12th November 2006, Berlin.

Interview 11: consultant hired by the Finance Senate for the BWB bidding process, 12 March 2007.

Interview 12: senior SPD politician in the Coalition Government, 18th April 2007, Berlin.

Interview 13: leading Vivendi executive in the RWE/Vivendi/Allianz bidding team, 23 April 2008.

Interview 14: civil servant working at the Finance Senate and involved in the privatisation process, 13 June 2007.

Interview 15: CDU politician working at the Finance Senate during the privatisation,
15 November 2007.

Interview 16: senior SPD parliamentarian involved in Parliamentary Party internal debates on privatisation, 29 November 2007.

Interview 17: Public Services Trade Union (ÖTV) representative during the privatisation process, 12 December 2007.

Interview 18: civil servant at the Finance Senate (Media and PR Department) during the privatisation process, 12 December 2007.

VS Forschung | VS Research

Neu im Programm Politik

Michaela Allgeier (Hrsg.)

Solidarität, Flexibilität, Selbsthilfe

Zur Modernität der Genossenschaftsidee

2011. 138 S. Br. EUR 39,95
ISBN 978-3-531-17598-0

Susanne von Hehl

Bildung, Betreuung und Erziehung als neue Aufgabe der Politik

Steuerungsaktivitäten in drei Bundesländern

2011. 406 S. (Familie und Familienwissenschaft) Br. EUR 49,95
ISBN 978-3-531-17850-9

Isabel Kneisler

Das italienische Parteiensystem im Wandel

2011. 289 S. Br. EUR 39,95
ISBN 978-3-531-17991-9

Frank Meerkamp

Die Quorenfrage im Volksgesetzgebungsverfahren

Bedeutung und Entwicklung

2011. 596 S. (Bürgergesellschaft und Demokratie Bd. 36) Br. EUR 39,95
ISBN 978-3-531-18064-9

Martin Schröder

Die Macht moralischer Argumente

Produktionsverlagerungen zwischen wirtschaftlichen Interessen und gesellschaftlicher Verantwortung

2011. 237 S. (Bürgergesellschaft und Demokratie Bd. 35) Br. EUR 39,95
ISBN 978-3-531-18058-8

Lilian Schwalb

Kreative Governance?

Public Private Partnerships in der lokalpolitischen Steuerung

2011. 301 S. (Bürgergesellschaft und Demokratie Bd. 37) Br. EUR 39,95
ISBN 978-3-531-18151-6

Kurt Beck / Jan Ziekow (Hrsg.)

Mehr Bürgerbeteiligung wagen

Wege zur Vitalisierung der Demokratie

2011. 214 S. Br. EUR 29,95
ISBN 978-3-531-17861-5

Erhältlich im Buchhandel oder beim Verlag.
Änderungen vorbehalten. Stand: Juli 2011.

www.vs-verlag.de

Abraham-Lincoln-Straße 46
65189 Wiesbaden
tel +49 (0)6221.345 - 4301
fax +49 (0)6221.345 - 4229